T0189476

Studies in Computational Intelligence

Volume 780

Series editor

Janusz Kacprzyk, Polish Academy of Sciences, Warsaw, Poland
e-mail: kacprzyk@ibspan.waw.pl

The series "Studies in Computational Intelligence" (SCI) publishes new developments and advances in the various areas of computational intelligence—quickly and with a high quality. The intent is to cover the theory, applications, and design methods of computational intelligence, as embedded in the fields of engineering, computer science, physics and life sciences, as well as the methodologies behind them. The series contains monographs, lecture notes and edited volumes in computational intelligence spanning the areas of neural networks, connectionist systems, genetic algorithms, evolutionary computation, artificial intelligence, cellular automata, self-organizing systems, soft computing, fuzzy systems, and hybrid intelligent systems. Of particular value to both the contributors and the readership are the short publication timeframe and the worldwide distribution, which enable both wide and rapid dissemination of research output.

More information about this series at http://www.springer.com/series/7092

Seyedali Mirjalili

Evolutionary Algorithms and Neural Networks

Theory and Applications

 Springer

Seyedali Mirjalili
Institute for Integrated and Intelligent
  Systems
Griffith University
Brisbane, QLD
Australia

ISSN 1860-949X   ISSN 1860-9503 (electronic)
Studies in Computational Intelligence
ISBN 978-3-030-06572-0   ISBN 978-3-319-93025-1   (eBook)
https://doi.org/10.1007/978-3-319-93025-1

Printed on acid-free paper

This Springer imprint is published by the registered company Springer International Publishing AG
part of Springer Nature
The registered company address is: Gewerbestrasse 11, 6330 Cham, Switzerland

To my mother and father

Preface

This book focuses on both theory and application of evolutionary algorithms and artificial neural networks. An attempt is made to make a bridge between these two fields with an emphasis on real-world applications.

Part I presents well-regarded and recent evolutionary algorithms and optimisation techniques. Quantitative and qualitative analyses of each algorithm are performed to understand the behaviour and investigate their potentials to be used in conjunction with artificial neural networks.

Part II reviews the literature of several types of artificial neural networks including feedforward neural networks, multi-layer perceptrons, and radial basis function network. It then proposes evolutionary version of these techniques in several chapters. Most of the challenges that have to be addressed when training artificial neural networks using evolutionary algorithms are discussed in detail.

Due to the simplicity of the proposed techniques and flexibility, readers from any field of study can employ them for classification, clustering, approximation, and prediction problems. In addition, the book demonstrates the application of the proposed algorithms in several fields, which shed lights to solve new problems. The book provides a tutorial on how to design, adapt, and evaluate artificial neural networks as well, which would be beneficial for the readers interested in developing learning algorithms for artificial neural networks.

Brisbane, Australia
April 2018

Dr. Seyedali Mirjalili

Contents

Acronyms

ABC	Artificial bee colony
ACO	Ant colony optimisation
ANN	Artificial neural network
AVE	Average
BAT	Bat algorithm
BBO	Biogeography-based optimisation
BP	Backpropagation
BPSO	Binary particle swarm optimisation
CS	Cuckoo search
DE	Differential evolution
DNN	Deep neural network
EA	Evolutionary algorithm
ES	Evolutionary strategy
FNN	Feedforward neural network
GA	Genetic algorithm
GBEST	Global best
GWO	Grey wolf optimiser
HSI	Habitat suitability index
ILS	Iterated local search
MLP	Multi-layer perceptron
MSE	Mean square error
N/A	Not application
NFL	No free lunch theorem
NN	Neural network
PBEST	Personal best

PBIL	Population-based incremental learning
PSO	Particle swarm optimisation
RBF	Radial basis function
SA	Simulated annealing
STD	Standard deviation

Part I
Evolutionary Algorithms

This part first provides preliminaries and essential definitions when optimising problems with single objective. The challenges involved when solving such problems are discussed as well. Several chapters are then given to present the inspirations and mathematical model of several evolutionary algorithms including:

- Particle swarm optimisation
- Ant colony optimisation
- Genetic algorithm
- Biogeography-based optimisation

Each of these algorithms is analysed theoretically and experimentally as well. For the experimental studies, the performance of algorithms is investigated when solving benchmark functions and real-world problems.

Chapter 1
Introduction to Evolutionary Single-Objective Optimisation

1.1 Introduction

In the past, the computational engineering design process used to be mostly experimentally based [1]. This meant that a real system first had to be designed and con-
structed to be able to do experiments. In other words, the design model was an actual
physical model. For instance, an actual airplane or prototype would have to put in a
massive wind tunnel to investigate the aerodynamics of the aircraft [2]. Obviously,
the process of design was very tedious, expensive, and slow.

After the development of computers, engineers started to simulate models in
computers to compute and investigate different aspects of real systems. This was a
revolutionary idea since there was no need for an actual model in the design phase
anymore. Another advantage of modelling problems in computers was the reduced
time and cost. For instance, it was no longer necessary to build a wind tunnel and
real model to compute and investigate the aerodynamics of an aircraft. The next step
was to investigate not only the known characteristics of the problem but also explore
and discover new features. Exploring the search space of the simulated model in a
computer allowed designers to better understand the problem and find optimal values
for design parameters. Despite the use of computer in modelling, a designer still had
to manipulate the parameters of the problem manually.

After the first two steps, people started to develop and utilise computational or
optimisation algorithms to use the computer itself to find optimal solutions of the
simulated model for a given problem. Thus, the computer manipulated and chose
the parameters with minimum human involvement. This was the birth of automated
and computer-aided design fields. Evolutionary Algorithms (EA) [3] also became
popular tools in finding the optimal solutions for optimisation problems.

Generally speaking, EAs mostly have very similar frameworks. They first start the
optimisation process by creating an initial set of random, trial solutions for a given
problem. This random set is then iteratively evaluated by objective function(s) of the
problem and evolved to minimise or maximise the objective(s). Although this framework is very simple, optimisation of real world problems requires considering and

© Springer International Publishing AG, part of Springer Nature 2019
S. Mirjalili, *Evolutionary Algorithms and Neural Networks*, Studies
in Computational Intelligence 780, https://doi.org/10.1007/978-3-319-93025-1_1

addressing several issues of which the most important ones are: local optima, expensive computational cost of function evaluations, constraints, multiple objectives, and uncertainties.

Real problems have mostly unknown search spaces that may contain many suboptimal solutions [4]. Stagnation in local optima is a very common phenomenon when using EAs. In this case, the algorithm is trapped in one of the local solutions and assumes it to be the global solution. Although the stochastic operators of EAs improve the local optima avoidance ability compared to deterministic mathematical optimisation approaches, local optima stagnation may occur in any EAs as well.

EAs are also mostly population-based paradigms. This means they iteratively evaluate and improve a set of solutions instead of a single solution. Although this improves the local optima avoidance as well, solving expensive problems with EAs is not feasible sometimes due to the need for a large number of function evaluations. In this case, different mechanisms should be designed to decrease the required number of function evaluations. Constraints are another difficulty of real problems [5], in which the search space may be divided into two regions: feasible and infeasible. The search agents of EAs should be equipped with suitable mechanisms to avoid all the infeasible regions and explore the feasible areas to find the feasible global optimum. Handling constraints requires specific mechanisms and has been a popular topic among researchers.

Real engineering problems often also have multiple objectives [6]. Optimisation in a multi-objective search space is quite different and needs special considerations compared to a single-objective search space. In a single-objective problem, there is only one objective function to be optimised and only one global solution to be found. However, in multi-objective problems there is no longer a single solution for the problem, and a set of solutions representing the trade-offs between the multiple objectives, the Pareto optimal set, must be found.

Last but not least, another key concept in the optimisation of real engineering problems is robustness [7]. Robust optimisation refers to the process of finding optimal solutions for a particular problem that have least variability in response to probable uncertainties. Uncertainties are unavoidable in the real world and can be classified in three categories: those affecting parameters, operating conditions, and outputs.

1.2   Single-Objective Optimisation

As its name implies, single-objective optimisation algorithms estimate the global optimum for single-objective problems. In a single-objective problem there is one set of optimal values for the parameters that leads to the most optimal objective value. The objective function is just one component of an optimisation problems. Other components are the inputs (parameters, variables) and constraints (See Fig. 1.1). The inputs are the unknowns form the optimiser perspective and produce one objective for every unique combination. The main duty of an optimisation algorithm is to find the optimal values for the inputs. Examples of the inputs are thickness of an airfoil, length of a propeller blade, diameter of a nozzle, etc.

Inputs   Constraints   Output

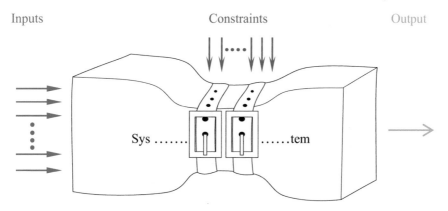

Fig. 1.1 Three main components of an optimisation system with single objective: inputs, constraints (operating conditions), and output

The constraints can be considered as secondary inputs for the system, but their impact is different from the primary inputs. The constraints indicate the limits of a system and should be considered to find feasible solutions. In fact, constraints define which set of values for the inputs are valid. The objective can be calculated for any set of inputs. However, a solution and its objective value is not acceptable in case of violation of the constraints. The optimisation algorithm do not find optimal values for constraints, yet it must consider them to find feasible solutions for decision makers.

Without the loss of generality, single-objective optimisation can be formulated as a minimisation problem as follows:

$$Minimise: \quad f(x_1, x_2, x_3, \ldots, x_{n-1}, x_n) \tag{1.1}$$

$$Subject\ to: \quad g_i(x_1, x_2, x_3, \ldots, x_{n-1}, x_n) \geq 0, i = 1, 2, \ldots, m \tag{1.2}$$

$$h_i(x_1, x_2, x_3, \ldots, x_{n-1}, x_n) = 0, i = 1, 2, \ldots, p \tag{1.3}$$

$$lb_i \leq x_i \leq ub_i, i = 1, 2, \ldots, n \tag{1.4}$$

where n is number of variables, m indicates the number of inequality constraints, p shows the number of equality constraints, lb_i is the lower bound of the i-th variable, and ub_i is the upper bound of the i-th variable.

This formulation shows that there are two types of constraints: equality and inequality. However, the equality constraints can be expressed as follow to formulate a single-objective problem with inequality constraints only:

$$h_i(x_1, x_2, x_3, \ldots, x_{n-1}, x_n) \leq 0 \land h_i(x_1, x_2, x_3, \ldots, x_{n-1}, x_n) \geq 0 \tag{1.5}$$

When formulating a problem, an optimiser would be able to tune its variables based on the outputs and constraints. As mentioned in the introduction of this chapter, one of the advantages of evolutionary algorithms is that they consider a system as a black

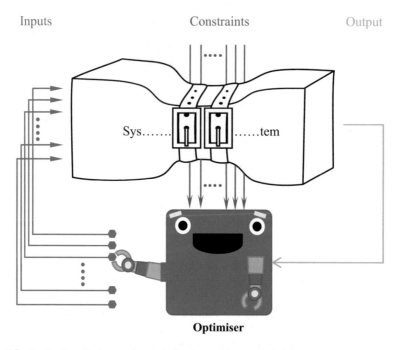

Inputs   Constraints   Output

Sys.......   ......tem

Optimiser

Fig. 1.2 Stochastic optimisers solve optimisation problem as a black box

box [8]. Figure 1.2 shows that the optimisers only provide the system with variables and observe the outputs. The optimisers then iteratively and stochastically change the inputs of the system based on the feedback (output) obtained so far until satisfaction of an end criterion. The process of changing the variables based on the history of outputs is defined by the mechanism of an algorithm. For instance, Particle Swarm Optimisation (PSO) [9] saves the best solutions obtained so far and encourages new solutions to relocate around them.

1.3 Search Landscape

The set of parameters creates a search space that will be searched by the optimiser. The set of parameters, objectives, and constraints create a search landscape. Figure 1.3 shows an example of search landscape with two variables and several constraints. Note that for most of the optimisation problems, the ranges are similar. Therefore, the search space is a square in 2D problems, a cube in 3D problems with three variables, and hyper-cube in problems with more than 3 variables. However, the range each of variable might be different. Therefore, the search space is no longer of cubic shape. In this case, we have to deal with a (hyper-)rectangular search space.

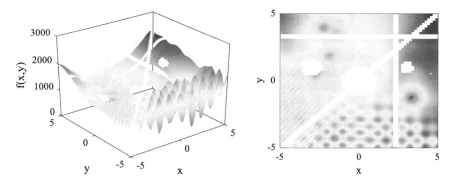

Fig. 1.3 A search landscape with several constraints

Depending on the relations between the input(s) and output(s), the search land scape can be unimodal or multi-modal. In a unimodal landscape, there is no local solutions (hills or valleys). In the multi-modal landscape, however, there are multiple local solutions of which one is the global optimum. It is worth mentioning here that the search landscape of real-world problems might be changing. In this case, a dynamic optimisation is required to keep track of the changing global optimum. Also, some search landscapes have multiple global optima with equal objective values.

Figure 1.3 shows that constraints create gaps in the search space that should be avoided by an optimisation algorithm. The constants may also divide the search land scape into multiple 'islands'. An optimisation algorithm should be able to discover isolated regions and identify the promising optima. It should be noted that each search landscape has boundaries, which can be considered as a set of constraints restricting the upper and lower bouts of the parameters. An optimisation algorithm should be equipped with suitable mechanisms to relocate over-shooted particles to the feasible regions of the search landscape. This search landscape illustrated in Fig. 1.3 is cal- culated using a set of simple mathematical functions. The landscape of real-world problems includes a large number of difficulties: a large number of parameters [10], a large number of constraints, a large number of local solutions, deceptive slopes, isolation of global optimum, dynamically changing optima, and uncertainties.

From an optimisation algorithm's perspective, the search landscape is defined by formulating the problem. This includes identifying and defining the number of parameters, range of each parameters, constraints, and the objectives. With the prob- lem formulation, an optimiser is able to find the optimal parameters to minimise or maximise the objectives. A conceptual model of an optimisation algorithm and its interaction with a system was shown in Fig. 1.2.

This figure shows that an optimisation algorithm (e.g. evolutionary algorithms) provides the inputs and monitors the outputs. The constraints are also considered to ensure the feasibility of the solutions during the optimisation. The way that an algorithm change the inputs is based on the structure of the algorithm. Optimisation algorithms normally require several iterations to determine the global optimum for a given problems.

1.4   Penalty Functions for Handling Constraints

It was mentioned in the preceding paragraph that the constraints are considered by the optimiser. In the literature, penalty functions are popular, in which a solution is penalized in case of violation of any of the constraints. One of the easiest penalty functions is called "barrier". In case of minimisation, an infeasible solution is penalized with a large objective value when violating the constraints in case of minimisation. As such, the algorithm ignores or improves this solution in the next iterations. An example of a barrier penalty function for the search landscape in Fig. 1.3 is given in Fig. 1.5. Note that the constraints applied to this search space are as follows:

$$(y \leq 3.2) \vee (y \geq 3.4) \tag{1.6}$$
$$(x \leq 2.2) \vee (x \geq 2.3) \tag{1.7}$$
$$(x - 3)^2 + (y - 1)^2 \geq 0.1 \tag{1.8}$$
$$(x + 3)^2 + (y - 1)^2 \geq 0.3 \tag{1.9}$$
$$x^2 + y^2 \geq 1 \tag{1.10}$$
$$x \neq y \tag{1.11}$$

The source code (written in Matlab) to implement the barrier function is shown in Fig. 1.4. It may be seen that if a solution violates a constraint at any level, it is penalised with a larger objective value (in case of minimisation). The impact of the barrier function on the shape of search landscape is shown in Fig. 1.5. It is evident that the infeasible regions became the biggest peaks, meaning that they are the worst regions of the search landscape. If a solution is generated on such regions by an optimisation algorithm, they have the poorest quality. Optimi-

Fig. 1.4 The code to implement the barrier penalty function. If a solution violates a constraints at any level, it is penalised with a larger objective value

```
if y > 3.2 & y < 3.4
    o = 2000;
end
if x > 2.2 & x < 2.3
    o = 2000;
end
if ((x-3)^2 + (y-1)^2) <0.1
    o = 2000;
end
if ((x+3) ^2 + (y-1)^2) <0.3
    o = 2000;
end
if (x ^2 + y^2) <1
    o = 2000;
end
if x == y
    o = 2000;
end
```

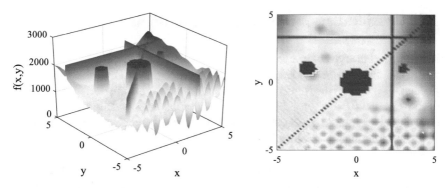

Fig. 1.5 A barrier penalty function applied to the constrained search landscape shown in Fig. 1.3

Fig. 1.6 A better way to penalise a solution is to add a large objective value to the current objective value returned for the infeasible solution

```
if y > 3.2 & y < 3.4
    o = f(x,y) + 2000;
end
if x > 2.2 & x < 2.3
    o = f(x,y) + 2000;
end
if ((x-3)^2 + (y-1)^2) <0.1
    o = f(x,y) + 2000;
end
if ((x+3) ^2 + (y-1)^2) <0.3
    o = f(x,y) + 2000;
end
if (x ^2 + y^2) <1
    o = f(x,y) + 2000;
end
if x == y
    o = f(x,y) + 2000;
end
```

sation algorithms normally ignore or significantly change such solutions to relocate them to better, feasible regions.

The issue with penalizing violated solutions with a equal value is that there is no differentiation between a solution that slightly violates one of the constrains or a solution that significantly violates multiple constraints. One way to address this issue is to add a large objective value to the value returned by the objective function for the infeasible solution. The code to implement this method is given in Fig. 1.6. It can be seen in this figure that the large objective value (2000) is now added to the objective value that is supported to return using $f(x, y)$. This results in having a non-flat regions for the violated regions. There are different works in the literature, so interested readers are refereed to [5, 11].

1.5 Classification of Optimisation Algorithms

Regardless of the type of search space and the penalty function, stochastic optimisation algorithms can be divided into two main classes [12]: individual-based versus population-based. In the former class, an optimisation algorithm starts with one candidate solution for a given problem. This solution is then evaluated and improved until the satisfaction of a termination condition. In the latter class, however, a set of candidate solutions is employed to determine the global optimum for optimisation problems.

The advantage of individual-based algorithm is the need for the minimum number of functions evaluations. In fact, the number of function evaluations is equal to the number of iterations. However, such techniques suffer from local optima stagnation (premature convergence). A single solution is very likely to trap in local solutions in real-world challenging problems.

By contrast, population-based techniques benefit from high exploration of search space and a lower probability of local optima entrapment. If a solution goes inside a local solution, other solutions will assist in avoiding it in the next iterations. As a drawback, such algorithms require more function evaluations and are computationally more expensive.

There is a large number of algorithms in each class. The most well-regarded individual-based algorithms are Simulated Annealing (SA) [13], Tabu search [14], Iterated Local Search (ILS) [15], and hill climbing [16].

TS is an improved local search technique that utilises short-term, intermediate-term, and long-term memories to ban and truncate unpromising/repeated solutions. Hill climbing is also another local search and individual-based technique that starts optimisation from a single solution. This algorithm then iteratively attempts to improve the solution by changing its variables. ILS is an improved hill climbing algorithm to decrease the probability of entrapment in local optima. In this algorithm, the optimum obtained at the end of each run is retained and considered as the starting point in the next run. Initially, the SA algorithm tends to accept worse solutions proportionally to a variable called the cooling factor. This assists SA to promote exploration of the search space and prevents it from becoming trapped in local optima when it does search them.

Over the course of last two decades, there have been many improvements in each of the aforementioned algorithms. The primary objective of most of them has been to alleviate the main drawback of such techniques: premature convergence. Despite several improvement [17] with successful applications [18], the nature of such techniques require them to show a lower exploration and local optima avoidance compared to population-based algorithms. A set of solutions is able to better handle the difficulties of real-world problems with a large number of local solutions.

1.6   Exploration and Exploitation

Population-based algorithms improve the set of initial solutions over the course of iterations. This is done in two phases: exploration (diversification) and exploitation (intensification) [19]. In the former phase, an algorithm abruptly changes the candidate solutions to ensure that they explore different regions of the search space. The main objective in the exploration phase is to discover the promising areas of the search landscape. In other words, a rough estimation of the global optimum for the problem is found in the exploration phase. The exploration can be done with frequent or large changes in the solutions. There are different operators for algorithms to provide exploratory behaviour. In PSO, for instance, the inertia weight maintains the tendency of particles toward their previous directions and emphasises exploration. In GA, a high probability of mutation causes random changes in individuals as the main mechanism for exploration.

In the exploitation phase, an algorithm requires the candidate solutions to search around the best solutions found so far. The magnitude of changes in the solutions is much lower than exploration. In other words, the search is done locally around the most promising solutions found in the exploration phase. The mechanism of exploitation is different depending on the type of an algorithm. In PSO, for instance, a low inertia rate causes low exploration and a higher tendency toward the best personal/global solutions obtained. Therefore, the particles converge toward best points instead of 'churning' around the search space. The mechanism that brings GA [20] exploitation is crossover. The crossover process causes slight random changes in the individuals and local search around the candidate solutions.

What makes the process of designing and utilising population-based algorithms difficult is that the exploration and exploitation are in conflict. A mere exploratory behaviour mostly results in finding poor-quality solutions since the algorithm never get the chance to improve the accuracy of the solutions. By contrast, a mere exploitative behaviour results in trapping in local solutions since they algorithm does not change the solutions abruptly. A good balance between exploration and exploitation lead to finding the global optimum for all types of problems. However, the challenge is that the intensity of these two conflicting phases is hard to define and balance. Finding a good balance between exploration and exploitation is difficult due ot the different search space of problems.

To solve problems with population-based algorithms, it is common to use adaptive operators to tune exploration and exploitation proportional to the number of iterations. This means that the algorithm smoothly transits from exploration to exploitation as the iteration counter increases. Another popular technique is to promote exploration at any stages of optimisation if there is no improvement in the best solution obtained so far (or all solutions). In PSO, for instance, the solutions can be changed randomly if the *pbest* and *gbest* do not change for a certain number of iterations.

1.7   Classification of Population-Based Optimisation Algorithms

Population-based algorithms can be divided into two main classes: evolutionary and swarm-based. Evolutionary algorithms mimic evolutionary concepts in nature to solve optimisation problems. They main operators in such techniques are cross-over and mutation. Cross-over components combine solutions during optimisation and are the main mechanism to exploit the search (local search). However, the mutation operators change some of the solutions significantly, which emphasizes exploration (global search). Some of the most popular proposed evolutionary algorithms are GA [21], DE [22], ES [23], and EP [24].

Swarm-based algorithms mostly define a position vector for each solutions and change (move) it using a set of rules. The movement step indicate direction and speed of solutions. Therefore, such techniques move solutions with the highest speed along different directions in a search space to perform exploration. The magnitude of movement reduces proportional to the number of iterations to exploit the best position(s) obtained in the exploration phase. Two of the most popular swarm-based algorithms are ACO [25] and PSO [26].

There is an important question here as to why stochastic algorithms have become significantly popular in different fields [27]. Firstly, the popularity of algorithms in the field of meta-heursitics is due the consideration of problem as a black box. Such techniques do not need derivation and are not based on gradient descent formulation. They provide the inputs to a problem and observe the output (see Fig. 1.2). In other words, stochastic algorithms are problem independent. Secondly, the inspiration of the algorithms is simple and easy to understand. The mechanisms are also understandable for researchers even without optimisation background. Thirdly, stochastic algorithms better handle difficulties of real-world problems compared to conventional optimisation techniques.

Another frequently asked question is as to why there is a substantial number of works despite the existence of many optimisation algorithms. There is a theory in the field of optimisation called No Free Lunch (NFL) [28] that logically proves there is no optimisation algorithm to solve all optimisation problems. This means that the solution to a particular problem is not necessarily the best algorithm on test functions or other real-world problems. To solve a problem one might need to modify, improve, or adapt an algorithm. This makes the field very active and is the reason why there is a large number of improvement of new algorithms every year.

Despite the advantages of meta-heuristics and their popularity, solving real-world problems requires special considerations. There is a large number of difficulties involved that should be considered to reliably find the global optimum.

1.8   Conclusion

This chapter provided preliminaries and essential definitions in the field of optimisation using meta-heuristics. Firstly, the challenges when solving single-objective
problems were covered briefly including the large number of variables, constraints,
multiple objectives, and uncertainties. Secondly, the main components of a single-
objective optimisation were discussed and formulated. The problems solved in this
book are single-objective, so this is the reason of providing an in-depth literature
review in this area. Thirdly, the concepts of search space and search landscape were
provided. Fourthly, one of the most popular methods for handling constraints were
give. Fifthly, the concepts of exploration and exploitation with examples in PSO and
GA were presented. Finally, the classification of individual-based and population-
based optimisation algorithms along with their pros and cons were explained.

The literature review of this chapter showed how popular EAs are for solving
challenging real-world problems. This does not mean that a designer can easily
solve a problem by just choosing an algorithm. An optimisation problem should
be formulated and prepared for an EA. This includes understanding the nature of
the problem, identifying the main components of the problem, and formulating the
objective function. Depending on the distinct difficulties and features of a particular
problem, an algorithm requires modifications. In training NNs, for instance, the
range of variables is unknown. Therefore, special mechanisms are needed to grow
the boundaries of the search landscape or require solutions to leave the search space
and discover new regions. Otherwise, a designer needs to make assumption for th
range of variables, which results in finding poor-quality solutions. This book is mainly
on EAs and their applications in training NNs. Therefore, such difficulties will be
discussed.

References

1. Klockgether, J., & Schwefel, H. P. (1970). Two-phase nozzle and hollow core jet experiments.
   In *Proceedings of 11th Symposium on Engineering Aspects of Magnetohydrodynamics* (pp.
   141–148). Pasadena, CA: California Institute of Technology.
2. NASA Ames National Full-Scale Aerodynamics Complex (NFAC). http://www.nasa.gov/
   centers/ames/multimedia/images/2005/nfac.html. Accessed 2016-08-16.
3. Hruschka, E. R., Campello, R. J., & Freitas, A. A. (2009). A survey of evolutionary algorithms
   for clustering. *IEEE Transactions on Systems, Man, and Cybernetics, Part C (Applications and
   Reviews)*, 39(2), 133–155.
4. Addis, B., Locatelli, M., & Schoen, F. (2005). Local optima smoothing for global optimization.
   Optimization Methods and Software, 20(4–5), 417–437.
5. Coello, C. A. C. (2002). Theoretical and numerical constraint-handling techniques used with
   evolutionary algorithms: a survey of the state of the art. *Computer methods in applied mechanics
   and engineering*, 191(11–12), 1245–1287.
6. Zhou, A., Qu, B. Y., Li, H., Zhao, S. Z., Suganthan, P. N., & Zhang, Q. (2011). Multiobjective
   evolutionary algorithms: A survey of the state of the art. *Swarm and Evolutionary Computation*,
   1(1), 32–49.

7. Mirjalili, S., Lewis, A., & Mostaghim, S. (2015). Confidence measure: a novel metric for robust meta-heuristic optimisation algorithms. *Information Sciences, 317*, 114–142.
8. Droste, S., Jansen, T., & Wegener, I. (2006). Upper and lower bounds for randomized search heuristics in black-box optimization. *Theory of computing systems, 39*(4), 525–544.
9. Shi, Y., & Eberhart, R. C. (1999). Empirical study of particle swarm optimization. In *Proceedings of the 1999 congress on evolutionary computation, CEC 99* (Vol. 3, pp. 1945–1950). IEEE.
10. Chu, W., Gao, X., & Sorooshian, S. (2011). Handling boundary constraints for particle swarm optimization in high-dimensional search space. *Information Sciences, 181*(20), 4569–4581.
11. Mezura-Montes, E., & Coello, C. A. C. (2006). A survey of constraint-handling techniques based on evolutionary multiobjective optimization. In *Workshop paper at PPSN*.
12. Mirjalili, S. (2016). SCA: a sine cosine algorithm for solving optimization problems. *Knowledge-Based Systems, 96*, 120–133.
13. Hwang, C. R. (1988). Simulated annealing: Theory and applications. *Acta Applicandae Mathematicae, 12*(1), 108–111.
14. Glover, F. (1989). Tabu searchpart I. *ORSA Journal on Computing, 1*(3), 190–206.
15. Loureno, H. R., Martin, O. C., & Stutzle, T. (2003). Iterated local search. International series in operations research and management science, 321–354.
16. Goldfeld, S. M., Quandt, R. E.,& Trotter, H. F. (1966). Maximization by quadratic hill-climbing. *Econometrica: Journal of the Econometric Society*, 541–551.
17. BoussaD, I., Lepagnot, J., & Siarry, P. (2013). A survey on optimization metaheuristics. *Information Sciences, 237*, 82–117.
18. Senvar, O., Turanoglu, E., & Kahraman, C. (2013). Usage of metaheuristics in engineering: A literature review. In *Meta-heuristics optimization algorithms in engineering, business, economics, and finance* (pp. 484–528). IGI Global.
19. repinek, M., Liu, S. H., & Mernik, M., (2013). Exploration and exploitation in evolutionary algorithms: A survey. *ACM Computing Surveys (CSUR), 45*(3), 35.
20. Holland, J. H. (1992). Genetic algorithms. *Scientific American, 267*(1).
21. Goldberg, D. E., & Holland, J. H. (1988). Genetic algorithms and machine learning. *Machine Learning, 3*(2), 95–99.
22. Das, S., & Suganthan, P. N. (2011). Differential evolution: A survey of the state-of-the-art. *IEEE Transactions on Evolutionary Computation, 15*(1), 4–31.
23. Mezura-Montes, E., & Coello, C. A. C. (2005). A simple multimembered evolution strategy to solve constrained optimization problems. *IEEE Transactions on Evolutionary Computation, 9*(1), 1–17.
24. Yao, X., & Liu, Y. (1996). *Fast evolutionary programming. Evolutionary programming, 3*, 451–460.
25. Dorigo, M., & Di Caro, G. (1999). Ant colony optimization: a new meta-heuristic. In *Proceedings of the 1999 congress on evolutionary computation, CEC 99* (Vol. 2, pp. 1470–1477). IEEE.
26. Kennedy, J., & Eberhart, R. (1995). Particle swarm optimization. In *Proceedings of IEEE international conference on neural networks* (Vol. 4, pp. 1942–1948). IEEE.
27. Dasgupta, D., & Michalewicz, Z. (Eds.). (2013). *Evolutionary algorithms in engineering applications*. Springer Science & Business Media.
28. Wolpert, D. H., & Macready, W. G. (1997). No free lunch theorems for optimization. *IEEE Transactions on Evolutionary Computation, 1*(1), 67–82.

Chapter 2
Particle Swarm Optimisation

2.1 Introduction

Particle Swarm Optimisation (PSO) algorithm [1] is one of the most well-regarded algorithm in the literature of stochastic optimisation approaches. It belongs to the family of swarm-based techniques and is a population-based algorithm. As GA [2] and ACO [3], it is of the most well-regarded algorithm in the literature. The application of this algorithm can be found in a large number of fields that shows its merits in solving problems of different types. There are different versions of this algorithm in the literature to solve constrained, dynamic, discrete, multi-objective, multi-modal, and many-objective problems. This chapter presents and analyses continuous and binary versions of this algorithm since in Part II of this book, both algorithms will be used in training NNs.

2.2 Inspiration

The PSO algorithm [1] simulates the navigation mechanism of birds in nature. The main inspiration is the simple equations proposed by Raynold in 1987 [4] for simulating the interaction of individuals in flying swarms. Raynold shows there are three primitive principles in bird swards: separation, alignment, and cohesion. The concepts of these three principles are illustrated in Fig. 2.1

It may be seen in this figure that the separation applies forces to prevent collision. In the alignment, the flying direction of an individual is adjusted based on its neighbours. In cohesion, a certain distance is maintained between the individuals to avoid isolation.

Separation (S), alignment (A), and Cohesion (C) are defined as follows [4]:

$$S = -\sum_{j=1}^{N} \left(X - X_j \right) \tag{2.1}$$

© Springer International Publishing AG, part of Springer Nature 2019
S. Mirjalili, *Evolutionary Algorithms and Neural Networks*, Studies
in Computational Intelligence 780, https://doi.org/10.1007/978-3-319-93025-1_2

Seperation	Alignment	Cohesion

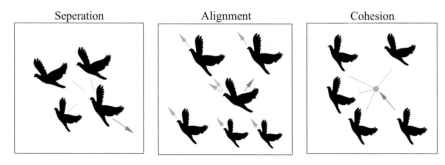

Fig. 2.1 Primitive corrective patterns between individuals in a bird swarm. In the alignment, the flying direction of an individual is adjusted based on its neighbours. In cohesion, a certain distance is maintained between the individuals to avoid isolation

where X shows the position of the current bird, X_j indicates the jth bird in the vicinity, and N is the number of birds in the neighbourhood.

$$A = \frac{\sum_{j=1}^{N} X_j}{N} \tag{2.2}$$

where X_j indicates the jth bird in the vicinity and N is the number of birds in the neighbourhood.

$$C = \frac{\sum_{j=1}^{N} X_j}{N} - X \tag{2.3}$$

where X shows the position of the current bird, X_j indicates the jth bird in the vicinity, and N is the number of birds in the neighbourhood.

To illustrate this behaviour, a simulation is done by implementing the above-mentioned equations. Note that in this simulation, a moving food source impacts the swarm as well and the velocity of each bird, which is defined by the following equation:

$$V_j(t+1) = V(t) + 0.3S + 0.3A + 0.3C + (X^* - X) \tag{2.4}$$

where $V_j(t)$ is the velocity of jth bird in the tth unit of time and X^* shows the position of a target or a source of food.

The position of each bird is updated using the following equation:

$$X(t+1) = X(t) + V(t+1); \tag{2.5}$$

The results are visualized in Fig. 2.2. This figure shows how the simple rules result in collision avoidance while maintaining the integrity of the entire swarm. This experiment shows how the birds adjust their positions, but it does not show the

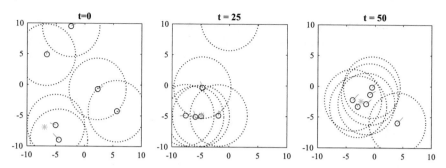

Fig. 2.2 Primitive corrective patterns between individuals in a bird swarm. In the first iteration, the position and velocity of each bird is random. Some birds form a small group after nearly 25 iterations, but the velocities are not aligned. A swarm is formed after nearly 50 iterations, in which the birds alight the velocities while maintain enough space to stay in the swarm and avoid collision

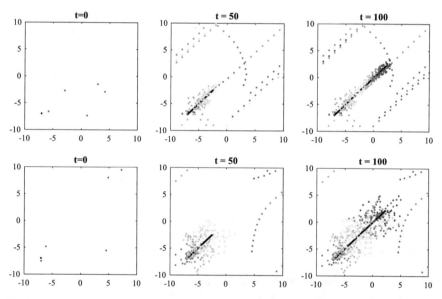

Fig. 2.3 Trajectory of five birds in a bird swarm (two independent simulations, first row: 5 particles, second row: 10 particles). The points with cool colours are generated in the initial iterations. The colour of points becomes warm proportional to the number of iterations

movement trajectory of each particle. To see how each bird moves in a 2D space, Fig. 2.3 is given. Note that in this simulation, five birds are moved and if they hit the boundary, they appear from opposite side of the search space. The radius of neighbourhood is equal to five as well.

It may be seen in the first simulation (first row in Fig. 2.3) that all birds chase the moving prey, while one particle just flies around. In this second simulation (second row in Fig. 2.3), however, all birds move around the prey. Gravitating the isolated bird towards the swarm is noticeable in both runs.

Due to the more number of birds moving around the prey in the second simulation, the dispersion of particles trajectory is higher than the first swarm. This is due to the separation component since birds are more likely to collide proportional to their number.

2.3 Mathematical Model of PSO

In the PSO algorithm every solution of a given problem is considered as a 'particle' which is able to move in a search landscape. In order to update the position of each particle, two vectors are considered: position vector and velocity vector. The position vector(X) defines the value for the parameters of the problem. This can be considered as the location of a particle in a d dimensional space where d is the number of variables. The velocity vector(V) defines direction and intensity of movement.

The location of particles is updated in each step of optimisation using the following equation [5]:

$$X_i(t+1) = X_i(t) + V_i(t+1) \tag{2.6}$$

where $X_i(t)$ shows the position of ith at tth iteration and $V_i(t)$ shows the velocity of ith at tth iteration.

This equation shows that the position updating is simple and the main component is the velocity vector. The velocity vector is defined as follows:

$$V_i(t+1) = wV_i(t) + c_1 r_1 \left(P_i(t) - x_i(t) \right) + c_2 r_2 \left(G(t) - x_i(t) \right) \tag{2.7}$$

where $X_i(t)$ shows the position of ith at tth iteration, $V_i(t)$ shows the velocity of ith at tth iteration, w is the inertia weight, c_1 shows the individual coefficient, c_2 signifies the social coefficient, r_1, r_2 are random vectors in $[0, 1]$, $P_i(t)$ is the best solution obtained by the ith particle until tth iteration, and $G(t)$ shows the best solution found by all particles (entire swarm) until ith iteration.

In this equation, three components are considered that slightly similar to the concepts of separation, alignment, and attraction discussed above. The first part $(wV_i(t))$ considers the current velocity which directly translate to the movement direction and speed. The parameter w, which is called inertia weight, indicates how much the previous movement direction and speed should be maintained. The second component is called cognitive component and simulates the individual intelligence of a particle. The $P(t)$ saves the best solution obtained so far by the particles and uses it as flag to update the position around the most promising regions of the search landscape.

The last part simulates the social intelligence of a swarm, in which each particle can see the best solution obtained so far by the entire swarm. Therefore, each particle gravitates towards the best regions of search space obtained so far.

In the PSO algorithm, the optimisation process starts with a set of random particles. Each particle is assigned with a position vector and velocity vector. The controlling parameters of PSO including w, c_1, and c_2 are initialised as well. The velocity vectors, position vectors, random components (r_1 and r_2), and the inertia weight (w) are repeatedly updated until the satisfaction of an end condition. The best solution obtained by the entire swarm is returned as the best estimation of the global optimum for the given problem.

What increases the chance of convergence of the PSO algorithm is the use of best solutions (including personal best and global best) when updating the position of particles. PSO maintains the best solutions obtained so far and requires the particles to search around them. This increases the probability of finding a better solution.

2.4 Analysis of PSO

As stated above, the velocity vector is the key element of the PSO algorithm. It consists of three components each of which plays an essential role in simulating the flocking behaviour of birds and solving optimisation problems. This section analyses each of these components in details.

The first component is $wV_i(t)$ which considers the current velocity of a particle when calculating the new velocity. The inertia weight defines how willing a particle is in maintaining its current velocity/direction. The inertia weight is a main controlling parameter in PSO to emphasize or balance exploration and exploitation. To see the impact of parameter, an experiment is conducted in the section, in which 20 particles are required to solve a multi-modal test functions during 100 number of iterations. The test functions is shown in Fig. 2.4.

This test functions is solved when $w = 0.9$, $w = 0.4$, and w linearly decreases from 0.9 to 0.4. The results are illustrated in Fig. 2.5. This figure shows that the particles highly move around the search space even in the final iterations (dots with warm colours) when the value of w is greater than 0.5. By contrast, the intensity

Fig. 2.4 Multi-modal test function (F9). The mathematical formulation of this function allows adding a desired number of local variables

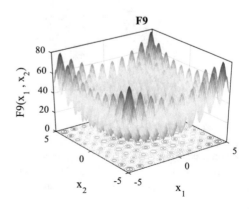

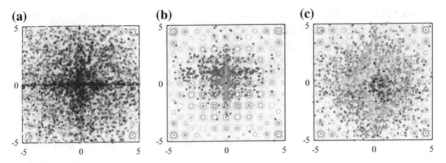

Fig. 2.5 The inertia weight is a main controlling parameter in PSO to emphasize or balance exploration and exploitation. History of particles' positions when **a** $w = 0.9$ **b** $w = 0.4$ **c** w linearly
decreases from 0.9 to 0.4. The particles highly move around the search space even in the final
iterations (dots with warm colours) when the value of w is hight. By contrast, the intensity of movement is substantially less when the inertia weight is less than 0.5. The last sub-plot shows that with
linearly decreasing the value of inertia weight, exploration and exploitation can be balanced

of movement is substantially less when the inertia weight is less than 0.5. The last
sub-plot in Fig. 2.5 shows that with linearly decreasing the value of inertia weight,
exploration and exploitation can be balanced. This means that the movement and
speed of particles decrease proportional to the number of iterations.

The social component saves the best solution obtained so far and highly favour
exploitation. A PSO with only social component is a hill-climbing algorithm. A
PSO with a cognitive component is an algorithm simulating multiple hill climbing
methods. This can be seen in Fig. 2.6. This figure shows that when the particles
only use the global best in the position updating, they quickly converge towards a

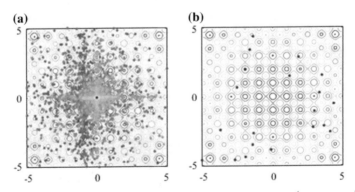

Fig. 2.6 Search history of particles when **a** $V_i(t + 1) = wV_i(t) + c_2r_2\left(G(t) - x_i(t)\right)$. When
the particles only use the global best in the position updating, they quickly converges towards
a point, which is normally a local optimal solution as the exploration is low. **b** $V_i(t + 1) =
wV_i(t) + c_1r_1\left(P_i(t) - x_i(t)\right)$. When the particles only use the personal bests, however, each par-
ticle finds a different solution and they all locally search one part of the search space

point, which is normally a local optimal solution as the exploration is low. When the particles only use the personal bests, however, each particle finds a different solution and they all locally search one part of the search space.

2.5 Standard PSO

The PSO presented in the preceding section is the original version, which is still very efficient. The literature shows that there is a significant number of improvements in this algorithm. The standard version of this algorithm is release every year. One of the latest standard PSO algorithms [6] use the following equations to search for global optima:

$$v_i^k(t) = \chi \left(v_i^k(t) + c_1 r_1 \left(P_i^k - X_i^k(t) \right) - c_2 r_2 \left((G^k - X_i^k(t)) \right) \right) \tag{2.8}$$

$$x_i^k(t+1) = x_i^k(t) + v_i^k(t) \tag{2.9}$$

where $x_i^k(t)$ indicates the kth parameter in the position vector of the ith particle at the tth iteration, G is the best solution obtained by the swarm in tth iteration, P_i shows the best solution that the ith particle found so far in t^{th} iteration, r_1, r_2 are random numbers generated using a uniform distribution in $[0, 1]$, c_1 is a constant for individual component, c_1 is a constant for social component, and χ is the constriction factor calculated using the following equation [7]:

$$\chi = \frac{2}{\left| 2 - \varphi - \sqrt{\varphi^2 - 4\varphi} \right|}, \varphi = c_1 + c_2 \tag{2.10}$$

where $c_1 = c_2 = 2.05$.

The impact of the constriction factor is similar to that of the inertia weight (w) in balancing exploration and exploitation. However, it has been proved that the constant value for this parameter can be more efficient than the adaptive inertia weight. To compare exploration and exploitation of both algorithm, Fig. 2.7 is given.

This figure shows that the exploration of the conventional PSO is higher than the standard PSO. The standard PSO shows much less exploration, yet the coverage of promising regions is good. It seems that the balance of exploration and exploitation is very efficient in the standard PSO. To see how fast both algorithms converge towards the global optimum, a 30-dimensional version of the F9 test function is solved and the convergence curves are illustrated in Fig. 2.8

(a)   **(b)**

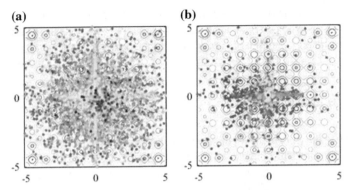

Fig. 2.7 a Conventional PSO versus **b** standard PSO. The exploration of the conventional PSO is higher than the standard PSO. The standard PSO shows much less exploration, yet the the coverage of promising regions is good. It seems that the balance of exploration and exploitation is very efficient in the standard PSO

Fig. 2.8 Comparison of conventional PSO (with ω) and standard PSO (with χ). The standard PSO converges towards the global optimum much faster than the conventional PSO

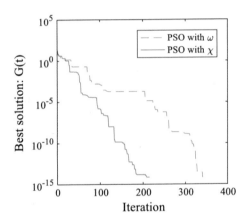

It may be seen in the standard PSO converges towards the global optimum much faster than the conventional PSO on the multi-modal test function. These results qualitatively show that the standard PSO is better than the conventional, despite the lower exploration. To fully investigate this, both algorithms are compared on 13 test functions and the results are given below. Note that the details of the test functions can be found in [8].

To solve the test functions, 60 particles and 1000 iterations are selected. Each algorithm is run 10 times and statistical results are given in Table 2.1. The Wilcoxon ranksum test is conducted in Table 2.2 at 5% significance level to confirm the significance of the results as well. To calculate the p-values, the algorithm with minimum *mean* in Table 2.2 is chosen and compared with other techniques. Since the best algorithm cannot be compared with itself, N/A is written in Table 2.2.

Table 2.1 Statistical results of conventional PSO and standard PSO

Function	Algorithm					
	Coventional PSO			Standard PSO		
	Mean	Std	Mediam	Mean	Std	Mediam
F1	2.4738E-07	2.3645E-07	1.4065E-07	9.6586E-20	2.0639E-19	9.1679E-21
F2	4.0002E+00	5.1639E+00	6.0200E-04	4.0000E+00	5.1640E+00	2.1900E-05
F3	1.3575E+01	3.1982E+00	1.4189E+01	4.7253E-01	3.0876E-01	4.9883E-01
F4	6.3693E-01	1.3663E-01	6.0735E-01	4.8617E-02	2.4707E-02	4.6593E-02
F5	5.0001E+01	4.4866E+01	2.7227E+01	3.0716E+01	2.2684E+01	2.2273E+01
F6	2.2800E-07	1.6900E-07	2.2400E-07	4.9100E-20	6.9000E-20	1.4900E-20
F7	4.0829E+00	4.2309E+00	2.7540E+00	1.2377E-02	4.1430E-03	1.2101E-02
F8	−6.5145E+03	6.4405E+02	−6.3599E+03	−6.9012E+03	8.1281E+02	−7.0713E+03
F9	7.3153E+01	2.0729E+01	7.7851E+01	8.5282E+01	2.2223E+01	8.5104E+01
F10	2.8100E-04	1.7900E-04	2.4200E-04	1.1552E-01	3.6529E-01	9.4700E-11
F11	1.1077E-02	9.6490E-03	9.8560E-03	1.8678E-02	1.7167E-02	1.6007E-02
F12	3.1300E-09	4.8000E-09	1.6900E-09	1.4520E-01	2.1981E-01	2.4500E-19
F13	2.1980E-03	4.6330E-03	6.5700E-08	3.2960E-03	5.3070E-03	9.6600E-21

Table 2.2 P-values obtained after conducting Wilcoxon ranksum test on the results in Table 2.1. P-values greate than or equal to 0.05 rejects the null hypothesis

Function	Algorithm	
	Coventional PSO	Standard PSO
F1	0.0002	N/A
F2	0.0539	N/A
F3	0.0002	N/A
F4	0.0002	N/A
F5	0.1405	N/A
F6	0.0002	N/A
F7	0.0002	N/A
F8	0.1859	N/A
F9	N/A	0.3847
F10	N/A	0.0028
F11	N/A	0.5702
F12	N/A	0.4727
F13	N/A	0.0538

It can be seen in the tables that the standard PSO outperforms the conventional PSO on all unimodal test functions as expected due to better exploitative ability of this algorithm. The p-vales in Table 2.2 indicate that the superiority is statistically significant on five of the unimodal test functions. This highly supports the statement about the better exploitation of the standard PSO. By contrast, the standard PSO provides very competitive results on multi-modal test functions. The p-values show that the conventional PSO is statistically better than the standard PSO on F10 test functions. These results indicate that the exploration of the conventional PSO is slightly superior.

2.6 Binary PSO

The original version of PSO is able to solve problems with continuous variables. The position and velocity vectors can be updated to any real numbers. One way to solve problems with discrete variables is to round real numbers to the nearest integer. In the literature, however, there is a systematic method to solve such problems with PSO. One of the most popular methods is to use a transfer function.

In the continuous PSO, the velocity indicates how fast a particle moves. If a particle is far from *pbest* and *gbest*, it faces sudden movements in the next iteration. In the binary PSO, a transfer function requires a particle to switch its position in a binary space using a probability calculated based on velocity as follows [9]:

$$T\left(V_i^k(t)\right) = \frac{1}{1 + e^{-v_i^k(t)}} \tag{2.11}$$

Fig. 2.9 The sigmoid transfer functions in BPSO

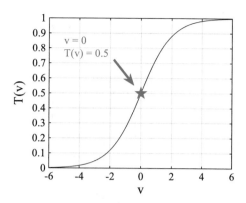

$$X_i^k(t + 1) = \begin{cases} 1 & r < T\left(V_i^k(t + 1)\right) \\ 0 & r \geq T\left(V_i^k(t + 1)\right) \end{cases} \tag{2.12}$$

where $v_i^k(t)$ indicates the kth parameter in the velocity vector of the ith particle at the tth iteration and r is a random number generated using a uniform distribution in $[0, 1]$.

The shape of this transfer function is shown in Fig. 2.9. It can be seen that the sigmoid transfer function returns 0.5 when the velocity is equation to 0. Eq. 2.12 indicates the in this case, the probability of changing a variable is at the highest level. This is a good mechanism to avoid local solution since the parameters can be assigned with 0 or 1 with equal probability.

By contrast, as the velocity values go toward positive infinity of negative infinity, the sigmoid transfer function returns values close to 1 or 0 respectively. This shows that the particle are moving away from the promising solutions (gbest and/or pbest), so it needs to change direction. This enforces the particle to get 0 or 1 for a given parameter as per Eq. 2.12.

The test functions have continuous parameters, so there should be a mechanism to convert discrete values to continuous. Similarly to [10], it is assumed that every particle includes a D bit string of length l each. Each bit is first converted to decimal values and then the following equation is used to bound it in the continuous interval of $[lb, ub]$:

$$X_{real} = lb + X_{decimal}\frac{ub - lb}{2^l} \tag{2.13}$$

where X_{real} is the continuous position vector, $X_{decimal}$ is the converted decimal number, and l is the length of each bit string (equals to 10 in this work).

Table 2.3 Statistical results of conventional BPSO and standard BPSO on continuous test functions

Function	Algorithm					
	Conventional BPSO			Standard BPSO		
	Mean	Std	Median	Mean	Std	Median
F1	5.4733E+02	1.4658E+02	5.0327E+02	9.1580E+02	1.7228E+02	9.6457E+02
F2	4.4727E+00	9.1486E-01	4.1113E+00	6.9828E+00	1.4090E+00	7.1973E+00
F3	9.2175E+02	1.8557E+02	9.0607E+02	1.0813E+03	3.0405E+02	1.1946E+03
F4	1.5273E+01	2.0971E+00	1.5625E+01	1.8770E+01	1.8562E+00	1.8359E+01
F5	3.7706E+04	1.6907E+04	3.1237E+04	7.1762E+04	3.8631E+04	5.4750E+04
F6	4.8294E+02	1.8498E+02	4.9351E+02	9.7874E+02	3.4153E+02	1.0478E+03
F7	9.0747E-02	4.8563E-02	7.5115E-02	1.2179E-01	5.7026E-02	1.1303E-01
F8	−6.5073E+03	3.1532E+02	−6.4800E+03	−6.2055E+03	3.2099E+02	−6.2441E+03
F9	2.8396E+01	4.4208E+00	2.8415E+01	3.7992E+01	4.4940E+00	3.9148E+01
F10	9.0429E+00	9.2259E-01	9.1044E+00	1.0951E+01	1.1339E+00	1.1393E+01
F11	4.9529E+00	1.7011E+00	5.0066E+00	9.5300E+00	2.4693E+00	9.8175E+00
F12	1.6157E+01	7.8399E+00	1.3328E+01	5.1812E+01	5.5346E+01	3.1795E+01
F13	3.4103E+04	3.0574E+04	3.1419E+04	3.8118E+05	3.3514E+05	2.8157E+05

Table 2.4 P-values obtained after conducting Wilcoxon ranksum test on the results in Table 2.3

Function	Algorithm	
	Coventional BPSO	Standard BPSO
F1	N/A	0.0010
F2	N/A	0.0017
F3	N/A	0.2123
F4	N/A	0.0015
F5	N/A	0.0140
F6	N/A	0.0036
F7	N/A	0.2413
F8	N/A	0.0890
F9	N/A	0.0008
F10	N/A	0.0036
F11	N/A	0.0008
F12	N/A	0.0058
F13	N/A	0.0002

To solve the test functions, 10-dimensional versions are chosen which require 100 binary values to solve. To estimate the global optimum, 60 particles and 1000 iterations are employed. As quantitative metrics, average, standard deviation, and mean of the best solution found by the algorithms at the end of 30 independent runs are collected. These two metrics show the overall performance and stability of algorithms. However, they do not consider each run independently. To consider each run and judge about the significance of the results, Wilcoxon ranksum test is conducted at 5% significance level ($p > 0.05$ rejects the null hypothesis) [11]. Note that all the test functions are unconstrained, so constraint handling is not considered when solving these case studies.

Both conventional and standard PSO algorithms are equipped with the sigmoid test functions, and the results are presented in Tables 2.3 and 2.4. Note that there are better transfer functions in the literature (e.g. v-shaped [12]). Finding the best transfer functions is outside the scope of this section since the main objective of this experiment is to compare the two most widely-used PSO algorithms under the same condition.

Inspecting the results of Table 2.3, it may be seen that the conventional BPSO significantly outperforms the standard PSO algorithm. Table 2.4 shows the this superiority is statistically significant on all test functions except F3, F7, and F8. The better results are possibly due to the higher exploration of the conventional BPSO.

The preceding test functions were continuous by nature and a mechanism was employed to convert them to discrete values. To compare the performance of the algorithms on real binary problems, a set of 0/1 knapsack problems. This problem is popular discrete problem with a large number of real-world applications such as task allocation [13], microgrid operation [14], scheduling hard real-time tasks [15], and financial modelling/planning [16]. This problem is formulated as follows:

$$\max \sum_{i=1}^{n} p_i x_i \tag{2.14}$$

$$s.t. : \sum_{i=1}^{n} r_i x_i \leq C \tag{2.15}$$

$$where : x_i \in [0, 1] \forall i \in [1, .., n] \tag{2.16}$$

where x_i is assigned one if the ith item is selected and zero otherwise. The paramter C is the maximum capacity (weight).

These equations show that the objective is to find a set of binary values for the vector x for maximising the profit of a selected set of items while the total weight does not exceed a maximum value. To create case studies, 25 datasets are extracted from an online resource,[1] which have been widely used [10, 17, 18].

In order to find the global optimum for these binary problems, 60 particles and 1000 iterations are used. The results are collected in 30 independent runs and presented in Tables 2.5 and 2.6. The results of both algorithms are identical on Knapsack problems with 8 and 12 variables. As the number of parameters raises, the results of the conventional BPSO become better. However, the p-values in Table 2.6 signify that the superiority is statistically significant occasionally. The reason of similar results of both algorithms is the low dimensionality of the Knapsack problems. In the test functions, there were 100 variables. In binary case studies employed in this section, however, there is a maximum of 24.

In summary, the results of this section shows that the BPSO algorithm is very efficient in solving binary problems. It was observed that the adaptive inertia weight assists BPSO to solve a wider range of problems.

---

[1] http://www.math.mtu.edu/~kreher/cages/Data.html.

Table 2.5 Comparison of conventional and standard BPSOs on 0/1 Knapsack problem

Name	n	Best known	Algorithm Conventional BPSO			Standard BPSO		
			Mean	Std	Median	Mean	Std	Median
Ks_8a	8	3.92E+06	3924400	0	3924400	3924400	0	3924400
Ks_8b	8	3.81E+06	3813669	0	3813669	3813669	0	3813669
Ks_8c	8	3.35E+06	3347452	0	3347452	3347452	0	3347452
Ks_8d	8	4.19E+06	4187707	0	4187707	4187707	0	4187707
Ks_8e	8	4.96E+06	4955555	0	4955555	4955555	0	4955555
Ks_12a	12	5.69E+06	5688887	0	5688887	5688887	0	5688887
Ks_12b	12	6.50E+06	6498597	0	6498597	6498597	0	6498597
Ks_12c	12	5.17E+06	5170626	0	5170626	5170626	0	5170626
Ks_12d	12	6.99E+06	6992404	0	6992404	6992404	0	6992404
Ks_12e	12	5.34E+06	5337472	0	5337472	5337472	0	5337472
Ks_16a	16	7.85E+06	7850983	0	7850983	7850983	0	7850983
Ks_16b	16	9.35E+06	9352998	0	9352998	9352998	0	9352998
Ks_16c	16	9.15E+06	9151147	0	9151147	9151147	0	9151147
Ks_16d	16	9.35E+06	9348889	0	9348889	9348482	2227.04	9348889
Ks_16e	16	7.77E+06	7767875	4725.568	7769117	7768496	3400.627	7769117
Ks_20a	20	1.07E+07	10725594	5537.687	10727049	10726005	4289.592	10727049
Ks_20b	20	9.82E+06	9814504	12965.91	9818261	9814725	9866.379	9818261
Ks_20c	20	1.07E+07	10713034	3390.293	10714023	10714023	0	10714023
Ks_20d	20	8.93E+06	8927564	6616.497	8929156	8929156	0	8929156
Ks_20e	20	9.36E+06	9357461	2209.9	9357969	9356954	3038.764	9357969
Ks_24a	24	1.35E+07	13539137	13772.13	13549094	13530602	19332.42	13524340
Ks_24b	24	1.22E+07	12223112	14316.11	12233713	12224916	14273.14	12233713
Ks_24c	24	1.24E+07	12442544	9360.942	12445379	12444141	5818.868	12445379
Ks_24d	24	1.18E+07	11805206	13160.12	11810051	11803534	14175.52	11810051
Ks_24e	24	1.39E+07	13934773.9	9628.644064	13940099	13932770.27	9657.965	13940099

Table 2.6 P-values obtained
from the Wilcoxon ranksum
test for the results on Table 2.5

| Name | n | Algorithm | |
		Conventional BPSO	Standard BPSO
Ks_8a	8	N/A	N/A
Ks_8b	8	N/A	N/A
Ks_8c	8	N/A	N/A
Ks_8d	8	N/A	N/A
Ks_8e	8	N/A	N/A
Ks_12a	12	N/A	N/A
Ks_12b	12	N/A	N/A
Ks_12c	12	N/A	N/A
Ks_12d	12	N/A	N/A
Ks_12e	12	N/A	N/A
Ks_16a	16	N/A	N/A
Ks_16b	16	N/A	N/A
Ks_16c	16	N/A	N/A
Ks_16d	16	N/A	0.3337
Ks_16e	16	0.5702	N/A
Ks_20a	20	0.9864	N/A
Ks_20b	20	0.7502	N/A
Ks_20c	20	0.0419	N/A
Ks_20d	20	0.1608	N/A
Ks_20e	20	N/A	0.1447
Ks_24a	24	N/A	0.0495
Ks_24b	24	0.5314	N/A
Ks_24c	24	0.9130	N/A
Ks_24d	24	N/A	0.3698
Ks_24e	24	N/A	0.2412

2.7   Conclusion

This chapter analysed the performance of PSO and BPSO algorithms on a variety
of test problems. It was observed that the inertia weight significantly impacts on the
exploration and exploitation. The adaptive values for this parameter degrades exploration, yet it is a good mechanism to balance exploratory and exploitive behaviour
of BPSO.

The conventional and standard PSO and BPSO were compared as well. It was
found that the conventional PSO and BPSO are better than the standard versions
when solving high-dimensional problems due to the high exploration. The PSO
algorithm is undoubtedly a well-regarded algorithm and will be applied to NNs in
the later chapters. The problem of training NNs changes for every dataset and has a

large number of locally optimal solutions. The conventional PSO algorithm will be applied to this problem due the a better balance of exploration and exploitation. The BPSO algorithm will be employed for feature selection in Chap. 9.

References

1. Eberhart, R., Kennedy, J. (1995). A new optimizer using particle swarm theory. In *1995 Proceedings of the sixth international symposium on micro machine and human science, MHS'95* (pp. 39–43). IEEE.
2. Goldberg, D. E., & Holland, J. H. (1988). Genetic algorithms and machine learning. *Machine Learning*, *3*(2), 95–99.
3. Dorigo, M., Birattari, M., & Stutzle, T. (2006). Ant colony optimization. *IEEE Computational Intelligence Magazine*, *1*(4), 28–39.
4. Reynolds, C. W. (1987). Flocks, herds and schools: A distributed behavioral model. *ACM SIGGRAPH Computer Graphics*, *21*(4), 25–34.
5. Kennedy, J. (2011). Particle swarm optimization. In *Encyclopedia of machine learning* (pp. 760–766). New York: Springer.
6. Bratton, D., Kennedy, J. (2007). Defining a standard for particle swarm optimization. In *2007 Swarm intelligence symposium, SIS 2007* (pp. 120–127). IEEE.
7. Eberhart, R. C., Shi, Y. (2000). Comparing inertia weights and constriction factors in particle swarm optimization. In *Proceedings of the 2000 congress on evolutionary computation* (Vol. 1, pp. 84–88). IEEE.
8. Mirjalili, S. (2016). SCA: A sine cosine algorithm for solving optimization problems. *Knowledge-Based Systems*, *96*, 120–133.
9. Kennedy, J., Eberhart, R. C. (1997). A discrete binary version of the particle swarm algorithm. In *1997 IEEE international conference on systems, man, and cybernetics*. Computational Cybernetics and Simulation (Vol. 5, pp. 4104–4108). IEEE.
10. Bansal, J. C., & Deep, K. (2012). A modified binary particle swarm optimization for Knapsack problems. *Applied Mathematics and Computation*, *218*(22), 11042–11061.
11. Derrac, J., Garca, S., Molina, D., & Herrera, F. (2011). A practical tutorial on the use of nonparametric statistical tests as a methodology for comparing evolutionary and swarm intelligence algorithms. *Swarm and Evolutionary Computation*, *1*(1), 3–18.
12. Mirjalili, S., & Lewis, A. (2013). S-shaped versus V-shaped transfer functions for binary particle swarm optimization. *Swarm and Evolutionary Computation*, *9*, 1–14.
13. Yang, J., Zhang, H., Ling, Y., Pan, C., & Sun, W. (2014). Task allocation for wireless sensor network using modified binary particle swarm optimization. *IEEE Sensors Journal*, *14*(3), 882–892.
14. Choi, S., Park, S., Kim, H. M. (2011). The Application of the 0-1 Knapsack problem to the load-shedding problem in microgrid operation. In *Control and automation, and energy system engineering* (pp. 227–234). Heidelberg: Springer.
15. Nawrocki, J., Complak, W., Baewicz, J., Kopczyska, S., & Makowiaki, M. (2009). The Knapsack-Lightening problem and its application to scheduling HRT tasks. *Bulletin of the Polish Academy of Sciences: Technical Sciences*, *57*(1), 71–77.
16. Bretthauer, K. M., & Shetty, B. (2002). The nonlinear Knapsack problem algorithms and applications. *European Journal of Operational Research*, *138*(3), 459–472.
17. Liu, J., Mei, Y., & Li, X. (2016). An analysis of the inertia weight parameter for binary particle swarm optimization. *IEEE Transactions on Evolutionary Computation*, *20*(5), 666–681.
18. Lee, C. Y., Lee, Z. J., Su, S. F. (2006, October). A new approach for solving 0/1 knapsack problem. In *2006 International conference on systems, man and cybernetics, SMC'06* (Vol. 4, pp. 3138–3143). IEEE.

Chapter 3
Ant Colony Optimisation

3.1 Introduction

Ant Colony Optimisation (ACO) [1, 2] is one of the well-known swarm intelligence techniques in the literature. The original version of algorithm is suitable for solving combinatorial optimisation problems (e.g. vehicle routing and scheduling). However, they have been lots of modifications in this algorithm that make it capable of solving a wide rage of problems these days. Despite the similarity of this algorithm to other swarm intelligence techniques in terms of employing a set of solutions and stochastic nature, the inspiration fo this algorithm is unique. This chapter presents the inspirations and mathematical models of this algorithm for solving both combinatorial and continuous problems.

3.2 Inspiration

The main inspiration of the ACO algorithm is the concept of stigmergy [3] in nature. Stigmergy refers to the manipulation of environment by biological organisms to communicate with each other. What makes this type of communication unique is the fact that individuals communicate indirectly. The communication is local as well, meaning that individuals should be in the vicinity of the manipulated area to access it [4].

In an ant colony, ants constantly look for food sources around the nest in random directions. It has been proven that once an ant finds a food source, it marks the path with pheromone. The amount of the pheromone highly depends on the quality and quantity of the food source. The more and the better the source of food, the stronger and concentrated pheromone is deposited. When other ants perceive the presence of the pheromone, they also follow the pheromone trail to reach the food source. After getting a portion of the food, ants carry them to the nest and mark their own path to

© Springer International Publishing AG, part of Springer Nature 2019
S. Mirjalili, *Evolutionary Algorithms and Neural Networks*, Studies
in Computational Intelligence 780, https://doi.org/10.1007/978-3-319-93025-1_3

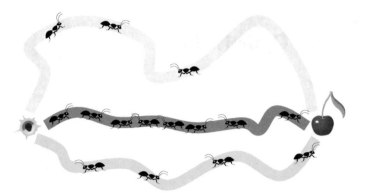

Fig. 3.1 Three routes from a nest to a food source. The amount of pheromone deposited on a route is of the highest on the closest path. The of pheromone decreases proportional to the length of the path. While ants add pheromone to the paths towards the food source, vaporization occurs. The period of time that an ant tops up pheromone before it vaporizes is higher inversely proportional to the length of the path. This means that the pheromone on the closets route become more concentrated as more ants get attracted to the strongest pheromone.

the next. It is interesting that this simple activity solves an optimisation problem for ant: finding the closest path from the nest to the food source.

While ants add pheromone to the paths towards the food source, vaporization occurs. However, the period of time that an ant tops up pheromone before it vaporizes is higher inversely proportional to the length of the path. This means that the pheromone on the closets route become more concentrated as more ants get attracted to the strongest pheromone. A conceptual example of this phenomenon is illustrated in Fig. 3.1.

The first theory of optimisation using ant models is called Ant System proposed in 1992 [5]. There are also other seminal works in this area such as Max-Min Ant System proposed in 1996 [6] and ACO proposed in 1997 [7]. The mathematical model and pseudo-code of the ACO algorithm is presented in the following sections.

3.3 Mathematical Model

As discussed above, the ACO algorithm was initially proposed to solve combinatorial problems [8]. In such problems, the objective is to find an optimal subset from a given finite set. Most of combinatorial problems are NP-hard, so heuristic approaches have been always beneficial and popular.

To solve combinatorial problems, ALO has been designed with three main phases: construction, pheromone update, and daemon (optional). The pseudo-code of the ACO algorithm and the sequence of the main phases are given in Fig. 3.2. The

Fig. 3.2 The pseudo code of
the ACO algorithm

```
Initialization
while the end condition is not satisfied
      Construction phase
      Pheromone update phase
      Optional daemon action phase
end
Return the best solution
```

concepts and mathematical models of construction, pheromone update, and daemon
phases are discussed in the following sections.

3.3.1 Construction Phase

In the construction phase, artificial ants are created from a finite set of n available
solutions [9]. Each ant can be seen as a set of values taken from the main finite set.
In the construction phase, a component from the main set is selected and added to
the artificial ant. The process is done based on the solution construction mechanism.

As an example, in the early Ant System, every ant was able to select an edge
in its current position when solving Traveling Salesman Problem (TSP). In each
iteration, an ant can chose all the unvisited edges available from the current node.
This selection process is done using the following probability:

$$p_{i,j} = \frac{(\tau_{i,j}^{\alpha})(\eta_{i,j}^{\beta})}{\sum(\tau_{i,j}^{\alpha})(\eta_{i,j}^{\beta})} \tag{3.1}$$

where $\tau_{i,j}$ shows the amount of pheromone of i, j edge, α defined the impact of
pheromone, $\eta_{i,j}$ indicates the desirability of i, j edge, and β defines the impact of the
desirability.

The desirability of an edge is defined by a weighting function. This function is
mostly heuristic and assigns a value which shows how good an edge is. If the distance
should be minimised, $\eta_{i,j}$ can be defined as follows:

$$\eta_{i,j} = \frac{1}{d_{i,j}} \tag{3.2}$$

where $d_{i,j}$ is the length of the edge i, j.

3.3.2 Pheromone Phase

This phase is the main mechanism to provide communication between the artificial ants and assist them in decision making for the construction phase [10]. The main components of this phase are pheromone evaporation and pheromone deposit. For solving the TSP problem, for example, the amount of pheromone on i, j edge is calculated as follows:

$$\tau_{i,j} = (1 - \rho)\tau_{i,j} + \Delta\tau_{i,j} \tag{3.3}$$

where ρ defines the rate of pheromone evaporation and $\Delta\tau_{i,j}$ is the total amount of pheromone deposited on the i, j edge defined as follows:

$$\Delta\tau_{i,j} = \begin{cases} \frac{1}{L_k} & k\text{th ant travels on edge } i, j \\ 0 & \text{otherwise} \end{cases} \tag{3.4}$$

where L_k shows the length (cost value) of the path that kth ant travels.

A general equation for the pheromone phase in which the pheromone values are defined by all ants (with a tour in TSP for instance) is written as follows:

$$\tau_{i,j} = (1 - \rho)\tau_{i,j} + \sum_{k=1}^{m} \Delta\tau_{i,j}^{k} \tag{3.5}$$

where m shows the number of ants, ρ is the rate of pheromone evaporation, and $\Delta\tau_{i,j}^{k}$ indicates the amount of pheromone deposited by the kth ant on edge i, j.

3.3.3 Daemon Phase

The ants in the ACO algorithm communicate locally using the phenomenon component. This phase adds a centralized control unit to assist ants in finding the best solution for a given optimisation problem. Different techniques can be used in this phase to bias the search. For instance, the best path in each iteration can be maintained or other local searches can be integrated to each artificial ant [11]. This step is very similar to elitism in evolutionary algorithms.

The original version of ACO (Ant System) was developed using the above-mentioned mathematical model. However. there is a lot of improvement afterwards. Three of them are discussed as follows.

3.3.4 Max-Min Ant System

In this version of Ant System [12], there is a range for the phenomenon concentration and the best ant is only allowed to lay phenomenon. In the Max-Min Ant System algorithm, the equation for updating pheromone is defined as follows:

$$\tau_{i,j} = \left[(1 - \rho)\tau_{i,j} + \Delta\tau_{i,j}^{best} \right]_{\tau_{lb}}^{\tau_{ub}} \tag{3.6}$$

where $\Delta\tau_{i,j}^{best}$ is the total amount of pheromone deposited on the i, j edge by the best ant, τ_{ub} is the upper bound of pheromone, τ_{lb} shows the lower bounds pheromone, and $[x]_{\tau_{lb}}^{\tau_{ub}}$ is defined as follows:

$$[x]_{\tau_{lb}}^{\tau_{ub}} = \begin{cases} ub & x > ub \\ lb & x < lb \\ x & \text{otherwise} \end{cases} \tag{3.7}$$

This component is similar to velocity clamping [13] in the PSO algorithm.

The summation of phronome deposited by the best ant is calculated as follow (note that the best ant can be selected from the set of best ants in the current iteration or the best ant obtained so far):

$$\Delta\tau_{i,j}^{best} = \begin{cases} \frac{1}{L_{best}} & \text{the best ant travels on edge } i, j \\ 0 & \text{otherwise} \end{cases} \tag{3.8}$$

where L_{best} shows the distance of the best path.

3.3.5 Ant Colony System

In the Ant Colony System [7], the focus was mostly on the Daemon phase, in which a mathematical model was employed to perform local pheromone update. In this approach, the following equation requires the ants to decrease the pheromone level so that the subsequent ants gravitate toward other paths:

$$\tau_{i,j} = (1 - \varphi)\tau_{i,j} + \varphi\tau_0 \tag{3.9}$$

where τ_0 shows that initial pheromone level and φ is a constant to reduce the pheromone level.

The process of decaying pheromone [14] by each ant promotes exploration of the search space since it reduces the probability of choosing the same path by other subsequent ants. Also, evaporation of pheromone prevents premature and rapid convergence.

3.3.6 Continuous Ant Colony

The easiest way to use ACO for solving continues problems is to split the variables' domain into a set of intervals [15]. This technique basically discretized the
continuous domain which results in low accuracy specially for problems with wide
domain variables. However, a combinatorial algorithm can be easily used without
modification. For instance, the domain of variable x in the interval of [0, 10] can be
discretized into $s = \{0, 1, 2, ..., 10\}$. The accuracy of this technique can be improved
with increasing the number of values in the set. Several continuous versions of ACO
have been developed in the literature to solve continuous problems without domain
modification. Interested readers are referred to [16, 17].

3.4 ACO for Combinatorial Optimisation Problems

As discussed above, the ACO algorithm is suitable for solving combinatorial problems. Since Knapsack problems were solved in the preceding chapters, a different
binary optimisation problem is chosen in this chapter to test the performance of the
ACO algorithm. Traveling Salesman Problem (TSP) problem [18] is one of the most
popular combinatorial problems with a wide range of applications in path planning,
routing, and resource allocation. It is considered as an NP hard problem, and most
of deterministic approaches are not effective when solving the high-dimensional
version of this problem.

In TSP, there are cities that should be visited by a sales man. There are different
variants of this problem in the literature. In the simplest version (symmetrical TSP),
the cost or distance between each pairs of city is known. The objective is to find
the shortest path from a starting point. A set of benchmark functions are extracted
from TSP Library (TSPLIB) [19] to observe the performance of the ACO algorithm
when solving such combinatorial optimisation problems. To solve the case studies,
40 ants, 300 iterations, $\alpha = 1$, $\beta = 1$, and $\rho = 0.5$ are used. The ACO algorithm is
run 30 times and three statistical metrics (mean, standard deviation, and median) are
reported in Table 3.1.

The ground truth of most of these case studies are unknown, so to see how the ACO
converges, some of the convergence curves are visualised in Figs. 3.3, 3.4 and 3.5.
All figures show that ACO benefits from a consistent convergence rate and improves
the first initial solution significantly over the course of iterations.

The TSPLIB test suit includes artificial and semi-real case studies. As the last
case study in this chapter, a real-world problem is solved in this chapter as well.
It is assumed that a tourist needs to visit 50 biggest cities in Australia. The main
objective is to find the shortest and cheapest path to travel the cities in Australia.
Another assumption is that the cost of travel is a function of distance. Figure 3.6
shows all the possible routes that a traveller might chose to see these 50 cities.

Table 3.1 The results of ACO on TSP case studies with a diverse number of cities

Test case	Number of cities	ACO		
		Mean	Std	Median
burma14	14	30.9446	0.1477	30.8785
bayg29	29	9614.5668	77.9339	9629.7404
att48	48	36425.5810	718.2049	36769.5119
eil51	51	491.4947	9.1905	489.2683
berlin52	52	8382.0477	131.6881	8383.5715
eil76	76	645.1189	16.1656	638.6704
kroA100	100	26054.8997	687.6009	26347.2171
lin105	105	17332.1160	174.0273	17324.4805
kroA150	150	34380.9286	618.8142	34504.7660
pr226	226	104465.2358	3425.8957	106052.0806
pr264	264	63309.1490	844.1970	63138.6712
pr299	299	70009.9046	1693.9420	70911.7764
lin318	318	60235.0933	709.4851	60224.6834
rd400	400	22693.5796	437.3995	22596.7226
pr439	439	156752.9542	3438.7965	157894.3520
att532	532	134648.0927	1040.1251	134706.0452
vm1084	1084	406811.7810	5509.8541	404039.5658

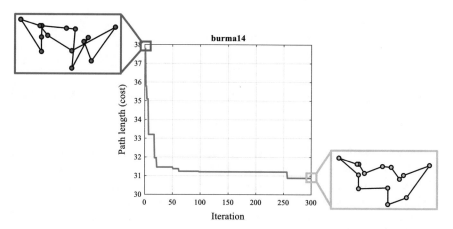

Fig. 3.3 The convergence of the ACO algorithm on the burma14 TSP case study (40 ants and 300 iterations)

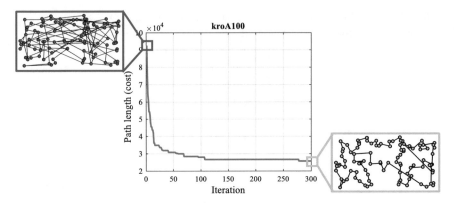

Fig. 3.4 The convergence of the ACO algorithm on the burma14 TSP case study (100 ants and 300 iterations)

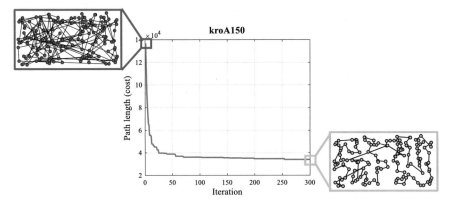

Fig. 3.5 The convergence of the ACO algorithm on the burma14 TSP case study (200 ants and 300 iterations)

To formulate this problem, the latitude and longitude of 50 biggest cities in Australia are used, and the ACO algorithm with 300 iterations and 40 ants are employed to solve this problem. The rest of the controlling parameters are identical to those used in the previous experiments. The best optimal path after 10 runs is visualised in Fig. 3.7. It may be seen that the path obtained is one of the shortest path from all possible paths in Fig. 3.6.

3.5   Conclusion

This chapter presented the inspirations and mathematical model of one of the most well-regarded swarm intelligence technique called ACO. Since ACO is a binary algorithm by nature, it was employed to solve a set of TSP problems. The results

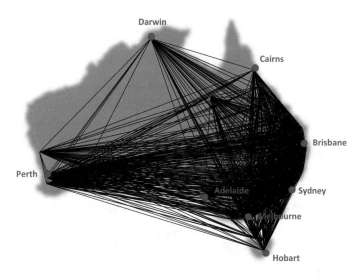

Fig. 3.6 All the possible routs between the 50 biggest cities in Australia

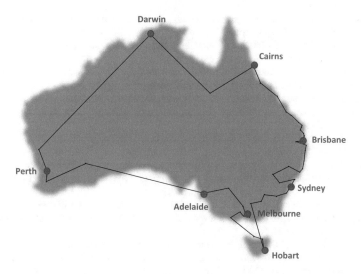

Fig. 3.7 The shortest path obtained by ACO to travel the 50 biggest cities in Australia

showed that this algorithm is very beneficial in solving combinatorial problems. For problems with continuous variables, however, the ACO algorithm should be equipped with mechanism to search in a continuous search space. This is essential for training NNs since the structural parameters of NNs are continuous.

References

1. Dorigo, M., & Birattari, M. (2011). Ant colony optimization. In *Encyclopedia of machine learning* (pp. 36–39). Boston, MA: Springer.
2. Dorigo, M., & Di Caro, G. (1999). Ant colony optimization: A new meta-heuristic. In *Proceedings of the 1999 Congress on Evolutionary Computation (CEC 99)* (Vol. 2, pp. 1470–1477). IEEE.
3. Grass, P. P. (1959). La reconstruction du nid et les coordinations interindividuelles chezBellicositermes natalensis etCubitermes sp. la thorie de la stigmergie: Essai d'interprtation du comportement des termites constructeurs. *Insectes sociaux, 6*(1), 41–80.
4. Dorigo, M., Bonabeau, E., & Theraulaz, G. (2000). Ant algorithms and stigmergy. *Future Generation Computer Systems, 16*(8), 851–871.
5. Dorigo, M., Maniezzo, V., & Colorni, A. (1996). Ant system: Optimization by a colony of cooperating agents. *IEEE Transactions on Systems, Man, and Cybernetics, Part B (Cybernetics), 26*(1), 29-41.
6. Sttzle, T., & Hoos, H. H. (1996). Improving the Ant System: A detailed report on the MAXMIN Ant System. FG Intellektik, FB Informatik, TU Darmstadt, Germany, Technical Report AIDA9612.
7. Dorigo, M., & Gambardella, L. M. (1997). Ant colony system: A cooperative learning approach to the traveling salesman problem. *IEEE Transactions on Evolutionary Computation, 1*(1), 53–66.
8. Papadimitriou, C. H., & Steiglitz, K. (1998). *Combinatorial optimization: Algorithms and complexity*. Courier Corporation.
9. Dorigo, M., & Sttzle, T. (2003). The ant colony optimization metaheuristic: Algorithms, applications, and advances. In *Handbook of metaheuristics* (pp. 250–285). Boston, MA: Springer.
10. Sttzle, T. (2009). Ant colony optimization. In *International Conference on Evolutionary Multi-Criterion Optimization* (pp. 2–2). Springer: Heidelberg.
11. Sttzle, T., Lpez-Ibnez, M., Pellegrini, P., Maur, M., De Oca, M. M., Birattari, M., & Dorigo, M. (2011). Parameter adaptation in ant colony optimization. In *Autonomous search* (pp. 191–215). Springer: Heidelberg.
12. Sttzle, T., & Hoos, H. H. (2000). MAXMIN ant system. *Future Generation Computer Systems, 16*(8), 889–914.
13. Shahzad, F., Baig, A. R., Masood, S., Kamran, M., & Naveed, N. (2009). Opposition-based particle swarm optimization with velocity clamping (OVCPSO). In *Advances in Computational Intelligence* (pp. 339–348). Springer: Heidelberg.
14. Sharvani, G. S., Ananth, A. G., & Rangaswamy, T. M. (2012). Analysis of different pheromone decay techniques for ACO based routing in ad hoc wireless networks. *International Journal of Computer Applications, 56*(2).
15. Socha, K. (2004). ACO for continuous and mixed-variable optimization. In *International Workshop on Ant Colony Optimization and Swarm Intelligence* (pp. 25–36). Springer: Heidelberg.
16. Socha, K., & Dorigo, M. (2008). Ant colony optimization for continuous domains. *European Journal of Operational Research, 185*(3), 1155–1173.
17. Blum, C. (2005). Ant colony optimization: Introduction and recent trends. *Physics of Life reviews, 2*(4), 353–373.
18. Hoffman, K. L., Padberg, M., & Rinaldi, G. (2013). Traveling salesman problem. In *Encyclopedia of operations research and management science* (pp. 1573–1578). Springer US.
19. Reinelt, G. (1991). TSPLIBA traveling salesman problem library. *ORSA Journal on Computing, 3*(4), 376–384.

Chapter 4
Genetic Algorithm

4.1 Introduction

Genetic Algorithm (GA) is one of the first population-based stochastic algorithm proposed in the history. Similar to other EAs, the main operators of GA are selection, crossover, and mutation. This chapter briefly presents this algorithm and applies it to several case studies to observe its performance.

4.2 Inspiration

GA was inspired from the Darwinian theory of evolutionary [1, 2], in which the survival of fitter creature and their genes were simulated. GA is a population-based algorithm. Every solution corresponds to a chromosome and each parameter represents a gene. GA evaluates the fitness of each individual in the population using a fitness (objective) function. For improving poor solutions, the best solutions are chosen randomly with a selection (e.g. roulette wheel) mechanism. This operator is more likely to choose the best solutions since the probability is proportional to the fitness (objective value). What increases local optima avoidance is the probability of choosing poor solutions as well. This means that if good solutions be trapped in a local solution, they can be pulled out with other solutions.

The GA algorithm is stochastic, so one might ask how reliable it is. What makes this algorithm reliable and able to estimate the global optimum for a given problem is the process of maintaining the best solutions in each generation and using them to improve other solutions. As such, the entire population becomes better generation by generation. The crossover between individuals results in exploiting the 'area' between the given two parent solutions. This algorithm also benefits from mutation. This operator randomly changes the genes in the chromosomes, which maintains the

© Springer International Publishing AG, part of Springer Nature 2019
S. Mirjalili, *Evolutionary Algorithms and Neural Networks*, Studies
in Computational Intelligence 780, https://doi.org/10.1007/978-3-319-93025-1_4

diversity of the individuals in the population and increases the exploratory behavior of GA. Similar to the nature, the mutation operator might result is a substantially better solution and lead other solutions towards the global optimum.

4.3 Initial Population

The GA algorithm starts with a random population. This population can be generated from a Gaussian random distribution to increase the diversity. This population includes multiple solutions, which represent chromosomes of individuals. Each chromosome has a set of variables, which simulates the genes. The main objective in the initialisation step is to spread the solutions around the search space as uniformly as possible to increase the diversity of population and have a better chance of finding promising regions. The next sections discuss the steps to improve the chromosomes in the first population.

4.4 Selection

Natural selection is the main inspiration of this component for the GA algorithm. In nature, the fittest individuals have a higher chance of getting food and mating. This causes their genes to contribute more in the production of the next generation of the same species. Inspiring from this simple idea, the GA algorithm employs a roulette wheel to assign probabilities to individuals and select them for creating the next generation proportional to their fitness (objective) values. Figure 4.1 illustrate an example of a roulette wheel for six individuals. The details of these individuals are presented in Table 4.1.

It can be seen that the best individual (#5) has the largest share of the roulette wheel, while the worst individual (#4) has the lowest share. This mechanism simulates the natural selection of the fittest individual in nature. Since a roulette wheel is a stochastic operator, poor individuals have a small probability of participating in the creation of the next generation. If a poor solution is 'lucky', its genes move to the next generation. Discarding such solutions will reduce the diversity of the population and should be avoided.

It should be noted that the roulette wheel is one of the many selection operators in the literature [3–5]. Some of the other selection operators are:

- Boltzmann selection [6]
- Tournament selection [7]
- Rank selection [8]
- Steady state selection [9]
- Truncation selection [10]
- Local selection [11]

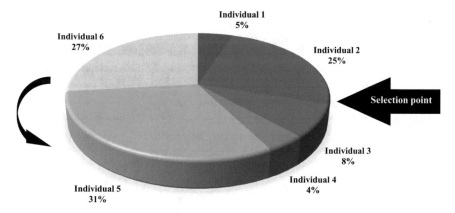

Fig. 4.1 Mechanism of the roulette wheel in GA. The best individual (#5) has the largest share of the roulette wheel, while the worst individual (#4) has the lowest share

Table 4.1 Details of the individuals in Fig. 4.1. The fittest individual is Individual #5

Individual number	Fitness value	% of Total
1	12	5
2	55	24
3	20	8
4	10	4
5	70	30
6	60	26
Total	227	100

- Fuzzy selection [12]
- Fitness uniform selection [13]
- Proportional selection [14]
- Linear rank selection [14]
- Steady-state reproduction [15]

4.5 Crossover (Recombination)

After selecting the individuals using a selection operator, they have to be employed to create the new generation. In nature, the chromosomes in the genes of a male and a female are combined to produce a new chromosome. This is simulated by combining two solutions (parent solutions) selected by the roulette wheel to produce two new solutions (children solutions) in the GA algorithm. There are different techniques for the crossover operator in the literature of which two (single-point and double-point [16]) are shown in Fig. 4.2.

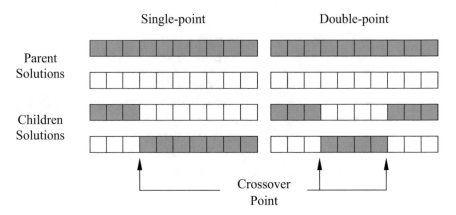

Fig. 4.2 Two popular crossover techniques in GA: single-point and double point. In the single-point cross over, the chromosomes of two parent solutions are swapped before and after a single point. In the double-point crossover, there are two cross over points and the chromosomes between the points are swapped only

In the single-point cross over, the chromosomes of two parent solutions are swapped before and after a single point. In the double-point crossover, however, there are two cross over points and the chromosomes between the points are swapped only. Other crossover techniques in the literature are:

- Uniform crossover [17]
- Half uniform crossover [18]
- Three parents crossover [19]
- Partially matched crossover [20]
- Cycle crossover [21]
- Order crossover [22]
- Position-based crossover [23]
- Heuristic cross over [24]
- Masked crossover [25]
- Multi-point crossover [26]

4.6   Mutation

The last evolutionary operator, in which one or multiple genes are altered after creating children solutions. The mutation rate is set to low in GA because high mutation rates convert GA to a primitive random search. The mutation operator maintains the diversity of population by introducing another level of randomness. In fact, this operator prevents solutions to become similar and increase the probability of avoiding local solutions in the GA algorithm. A conceptual example of this operator

Fig. 4.3 Mutation operator alters one or multiple genes in the children solutions after the crossover phase

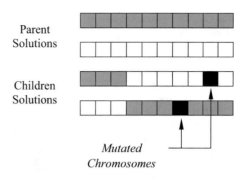

Parent Solutions

Children Solutions

Mutated Chromosomes

is visualised in Fig. 4.3. It can be seen in this figure that slight changes in the some of the randomly selected genes occur after the crossover (recombination) phase.

Some of the popular mutation techniques in the literature are:

- Power mutation [27]
- Uniform [28]
- Non-uniform [29]
- Gaussian [30]
- Shrink [31]
- Supervised mutation [32]
- Uniqueness mutation [33]
- Varying probability of mutation [34]

Taken together, most of EAs use the three evolutionary operators: selection, crossover, and mutation. These operators are applied to each generation to improve the quality of genes in the next generation. Another popular evolutionary operator is elitism [35], in which one or multiple best solutions are maintained and transferred without modification to the next generation. The main objective is to prevent such solutions (elites) from being degraded when applying the crossover or mutation operators.

The GA algorithm starts with a random population of individuals. Until the end of the end criterion, this algorithm improves the population using the above-mentioned three operators. The best solution in the last population is returned as the best approximation of the global optimum for a given problem. The rate of selection, crossover, and mutation can be changed or set to fix numbers during the optimisation. The next sections investigate the impact of changing such rates on the performance of GA.

4.7 Experiments When Changing the Mutation Rate

In this section several experiments are conducted to observe the impact of crossover and mutation rates on the performance of the GA algorithm. In the first experiment,

Fig. 4.4 Average objective value of 30 runs on all test functions for GA1 to GA6. The performance of the GA can be improved when the mutation is increased up to 0.4–0.6. However, this algorithm shows the worst average results when the mutation is very high. This is because the randomness of the search process increases proportional to the mutation rate

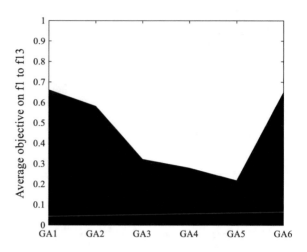

the mutation rate of GA is assigned with 0, 0.1, 0.2, 0.4, 0.6, 0.8 numbers to design GA1, GA2, GA3, GA4, GA5, and GA6 algorithms respectively. Note that the GA algorithm generates a random number in the interval of [0, 1], and if this number less than the mutation rate (if $r \leq \mu$ where r is the random number and μ is the mutation rate), the gene is mutated. This process is repeated for each chromosome and gene.

To compare the performance of GA when using different mutation rates, the function F1–F13 ([36]) is solved 30 independent runs and the statistical result are given in Table 4.2. To see how significantly the results of algorithms differ, Table 4.3 present the results of the Wilcoxon ranksum test conducted at 5% significance level.

To visually compare the overall performance of the GA algorithms, the average objective value of the solutions on all test functions are calculated and visualized in Fig. 4.4. It can be seen in the figure that the performance of the GA can be improved when the mutation is increased up to 0.4–0.6. However, this algorithm shows the worst average results when the mutation rate is very high. This is because the randomness of the search process increases proportional to the mutation rate. A lot of changes in the individuals, when the mutation rate is high, results in primitive random search and minimising the impact of selection and crossover operators. To observe the search pattern, Fig. 4.5 illustrates the search history (in the first two dimensions) of GA with 40 individuals and 5000 generations when solving the $F1$ test function.

This figure shows that the search history is sparse when the mutation is low. As a consequence, the accuracy and local search is high since parent solutions constantly share genes with a small level of mutations. By contrast, the coverage of landscape is high in the right subplot of Fig. 4.5. This shows that the solutions face changes a lot when the mutation rate is high that results in a more randomized search.

Table 4.2 There results of algorithms when changing the mutation rate. In this experiment the number of individuals is 40, and the maximum number of iterations is 500

Test function	GA1		GA2		GA3		GA4		GA5		GA6	
	mean	std	mean	std	mean	std	mean	std	mean	std	mean	std
F1	0.1366	0.4803	1.0000	1.0000	0.9636	0.1620	0.0000	0.6728	0.1394	0.0000	0.7018	0.8134
F2	1.0000	1.0000	0.1397	0.1203	0.4333	0.5596	0.0000	0.0000	0.0274	0.0226	0.5248	0.4507
F3	1.0000	1.0000	0.1300	0.0583	0.0000	0.1164	0.3640	0.0000	0.2171	0.0117	0.1336	0.4866
F4	1.0000	1.0000	0.6840	0.4524	0.2991	0.0913	0.0000	0.0000	0.4442	0.0405	0.4277	0.6498
F5	1.0000	1.0000	0.5890	0.9196	0.0000	0.3892	0.2202	0.5494	0.0853	0.0000	0.7966	0.3292
F6	1.0000	1.0000	0.1603	0.2800	0.0000	0.5967	0.3006	0.2959	0.4282	0.0000	0.4053	0.4448
F7	0.8532	0.6479	1.0000	0.7878	0.6110	0.0871	0.6421	0.1483	0.0000	0.0000	0.6544	1.0000
F8	1.0000	0.2423	0.7861	0.0000	0.1699	1.0000	0.4122	0.2979	0.0000	0.5469	0.3323	0.7173
F9	0.1404	0.7163	0.0000	1.0000	0.5839	0.6126	0.4022	0.2895	0.1289	0.4869	1.0000	0.0000
F10	1.0000	0.2586	0.2338	0.7589	0.0187	0.7901	0.0000	0.8643	0.2661	1.0000	0.9602	0.0000
F11	0.2344	0.4171	0.8521	1.0000	0.6126	0.4757	0.6601	0.3436	0.0000	0.2859	1.0000	0.0000
F12	0.2457	0.2652	1.0000	0.0000	0.0000	0.2413	0.0303	0.7846	0.8027	0.3148	0.7802	1.0000
F13	0.0000	0.4795	1.0000	0.3164	0.4995	1.0000	0.6068	0.1053	0.2952	0.5249	0.7392	0.0000

Table 4.3 P-values obtained after conducting the Wilcoxon ranksum text at 5% significance level for the algorithms in Table 4.2

Test function	GA1	GA2	GA3	GA4	GA5	GA6
F1	0.8501	0.2413	0.2413	N/A	0.7337	0.7337
F2	0.6232	0.6232	0.2413	N/A	0.4727	0.3447
F3	0.0452	0.5205	N/A	0.2730	0.5205	0.6232
F4	0.1041	0.3447	0.5708	N/A	0.6776	0.2123
F5	0.3447	0.4274	N/A	0.7913	0.8501	0.3447
F6	0.0257	0.5708	N/A	0.5708	0.2730	0.4274
F7	0.0058	0.0046	0.0113	0.0073	N/A	0.0376
F8	0.0173	0.0539	0.6232	0.2123	N/A	0.5205
F9	0.7913	N/A	0.9097	0.9097	0.6776	0.9698
F10	0.1405	0.9698	0.7913	N/A	0.9698	0.1212
F11	0.9698	0.4727	0.3847	0.4727	N/A	0.0757
F12	0.9698	0.0376	N/A	0.9097	0.1405	0.3847
F13	N/A	0.0312	0.2730	0.1041	0.2413	0.0757

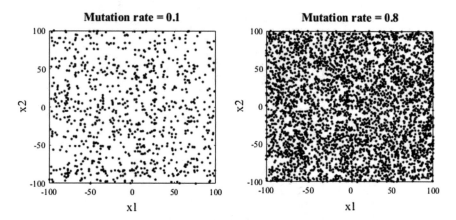

Fig. 4.5 Search history (in the first two dimensions) of GA with 40 individuals and 5000 generations when solving the F1 test function. The search history is sparse when the mutation is low. As the consequence, the accuracy and local search is high since parent solutions constantly share genes with a small level of mutations. By contrast, the coverage of landscape is high in the right subplot. This shows that the solutions face changes a lot when the mutation rate is high that results in a more randomized search

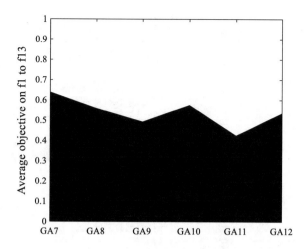

Fig. 4.6 Average objective value of 30 runs on all test functions for GA7 to GA12. The performance of GA does not degrade significantly when changing the crossover rate as opposed to the results in Fig. 4.4 when changing the mutation rate. The performance of GA is at best when the crossover rate is 0.9 (GA11)

4.8   Experiment When Changing the Crossover Rate

In the preceding section, the impact of the mutation rate on the performance of GA was investigated. In this section, the same experiments are conducted when changing the crossover rate: GA7 (0.5), GA8 (0.6), GA9 (0.7), GA10 (0.8), GA11 (0.9), GA12 (1). The average and standard deviation of 30 independent runs on F1–F13 are presented in Table 4.4. The p-values and average performance of GA on all test functions are shown in Table 4.5 and Fig. 4.6.

The results show that the performance of GA does not degrade significantly when changing the crossover rate as opposed to the results in the preceding section. Figure 4.6 shows that the performance of GA is at best when the crossover rate is 0.9 (GA11). High crossover rates allow creating more diverse children. As the main search mechanism, the crossover operator combines the best solutions to produce children. The higher and more frequently crossover, the higher probability of finding better children.

4.9   Conclusion

This section presented GA as one of the most popular evolutionary algorithm in the literature. After discussing the main evolutionary operators, several experiments were conducted to see the impact of changing controlling parameters on the performance of this algorithm. It was observed that the mutation rate should be tuned carefully since it can degrade the performance significantly when its value is high. It was also observed that the crossover rate is important in getting better results, but it can be set

Table 4.4 Experimental results when changing crossover rate: GA7 (0.5), GA8 (0.6), GA9 (0.7), GA10 (0.8), GA11 (0.9), GA12 (1). In this experiment the mutation was set to 0.6, the number of individuals are 40, and the maximum number of iterations are 500

Test function	GA7		GA8		GA9		GA10		GA11		GA12	
	mean	std	mean	std	mean	std	mean	std	mean	std	mean	std
F1	0.7508	0.7170	0.5297	1.0000	0.6734	0.4206	0.0000	0.0000	0.4046	0.5895	1.0000	0.0906
F2	1.0000	0.5773	0.4589	0.3578	0.5933	1.0000	0.0876	0.1178	0.0000	0.0000	0.1931	0.2154
F3	0.3315	0.0465	0.0569	1.0000	0.8908	0.4362	0.2770	0.6940	1.0000	0.3331	0.0000	0.0000
F4	0.0587	0.2214	0.9924	0.0000	0.5088	0.7690	0.9855	1.0000	0.0000	0.4858	1.0000	0.8390
F5	1.0000	0.4663	0.8640	0.0000	0.5640	0.1046	0.8677	0.6451	0.2526	1.0000	0.0000	0.7846
F6	0.8370	0.0601	0.8832	0.2771	0.3101	0.1238	1.0000	0.0000	0.3174	0.3928	0.0000	1.0000
F7	1.0000	0.0000	0.8933	0.3547	0.5416	0.3955	0.7573	0.2801	0.0000	0.5880	0.4453	1.0000
F8	0.6912	0.4536	0.0000	1.0000	0.7570	0.0000	0.5009	0.2983	0.3756	0.4908	1.0000	0.1683
F9	0.0945	1.0000	0.5832	0.7715	0.2822	0.0000	0.0000	0.8047	0.8402	0.8329	1.0000	0.5416
F10	0.4755	0.5376	0.8069	0.3889	0.9182	0.6203	1.0000	0.0000	0.0000	1.0000	0.6171	0.1449
F11	0.7502	0.3310	0.5221	0.2832	0.0665	0.0341	1.0000	0.0000	0.4164	1.0000	0.0000	0.7802
F12	0.4189	0.2010	0.0000	0.8621	0.2993	0.0000	0.5716	0.3662	0.9151	1.0000	1.0000	0.4007
F13	0.8967	0.0000	0.6746	0.0513	0.0000	0.7526	0.4259	0.3592	1.0000	1.0000	0.7011	0.8903

Table 4.5 P-values obtained after conducting the Wilcoxon ranksum text at 5% significance level for the algorithms in Table 4.4

Test function	GA7	GA8	GA9	GA10	GA11	GA12
F1	0.0257	0.3447	0.064	N/A	0.3075	0.0058
F2	0.0452	0.273	0.9698	0.9097	N/A	0.9097
F3	0.4727	0.6776	0.064	0.5708	0.0173	N/A
F4	0.7913	0.089	0.8501	0.1859	N/A	0.2123
F5	0.064	0.3447	0.4274	0.162	0.8501	N/A
F6	0.1041	0.1405	0.9097	0.0312	0.7337	N/A
F7	0.1041	0.273	0.4727	0.4727	N/A	0.7337
F8	0.1212	N/A	0.5205	0.7913	0.9698	0.1041
F9	0.9698	0.3447	0.7337	N/A	0.273	0.1212
F10	0.9698	0.5708	0.4274	0.7337	N/A	N/A
F11	0.3847	0.5205	0.9097	0.273	0.6232	N/A
F12	0.2123	N/A	0.2413	0.273	0.1859	0.0539
F13	0.064	0.1859	N/A	0.4274	0.089	0.2123

to high values without negative consequences as opposed to the mutation rate. For training NNs, a continuous version of GA should be employed as discussed in the preceding chapter (ACO). Due to the difficulty of training NN, both crossover and mutation rates will tuned.

References

1. Goldberg, D. E., & Holland, J. H. (1988). Genetic algorithms and machine learning. *Machine Learning, 3*(2), 95–99.
2. Holland, J. H. (1992). Genetic algorithms. *Scientific American, 267*(1), 66–73.
3. Genlin, J. (2004). Survey on genetic algorithm. *Computer Applications and Software, 2,* 69–73.
4. Cant-Paz, E. (1998). A survey of parallel genetic algorithms. *Calculateurs Paralleles, Reseaux et Systems Repartis, 10*(2), 141–171.
5. Goldberg, D. E., & Deb, K. (1991). A comparative analysis of selection schemes used in genetic algorithms. In *Foundations of genetic algorithms* (Vol. 1, pp. 69–93). Elsevier.
6. Goldberg, D. E. (1990). A note on Boltzmann tournament selection for genetic algorithms and population-oriented simulated annealing. *Complex Systems, 4*(4), 445–460.
7. Miller, B. L., & Goldberg, D. E. (1995). Genetic algorithms, tournament selection, and the effects of noise. *Complex Systems, 9*(3), 193–212.
8. Kumar, R. (2012). Blending roulette wheel selection & rank selection in genetic algorithms. *International Journal of Machine Learning and Computing, 2*(4), 365.
9. Syswerda, G. (1991). A study of reproduction in generational and steady-state genetic algorithms. In *Foundations of genetic algorithms* (Vol. 1, pp. 94–101). Elsevier.
10. Blickle, T., & Thiele, L. (1996). A comparison of selection schemes used in evolutionary algorithms. *Evolutionary Computation, 4*(4), 361–394.

11. Collins, R. J., & Jefferson, D. R. (1991). *Selection in massively parallel genetic algorithms* (pp. 249–256). University of California (Los Angeles), Computer Science Department.
12. Ishibuchi, H., & Yamamoto, T. (2004). Fuzzy rule selection by multi-objective genetic local search algorithms and rule evaluation measures in data mining. *Fuzzy Sets and Systems, 141*(1), 59–88.
13. Hutter, M. (2002). Fitness uniform selection to preserve genetic diversity. In *Proceedings of the 2002 Congress on Evolutionary Computation, CEC'02* (Vol. 1, pp. 783–788). IEEE.
14. Grefenstette, J. J. (1989). How genetic algorithms work: A critical look at implicit parallelism. In *Proceedings of the 3rd International Joint Conference on Genetic Algorithms (ICGA89)*.
15. Syswerda, G. (1989). Uniform crossover in genetic algorithms. In *Proceedings of the Third International Conference on Genetic Algorithms* (pp. 2–9). Morgan Kaufmann Publishers.
16. Srinivas, M., & Patnaik, L. M. (1994). Genetic algorithms: A survey. *Computer, 27*(6), 17–26.
17. Semenkin, E., & Semenkina, M. (2012). Self-configuring genetic algorithm with modified uniform crossover operator. In *International Conference in Swarm Intelligence* (pp. 414–421). Heidelberg: Springer.
18. Hu, X. B., & Di Paolo, E. (2007). An efficient genetic algorithm with uniform crossover for the multi-objective airport gate assignment problem. In *IEEE Congress on Evolutionary Computation, 2007 (CEC 2007)* (pp. 55–62). IEEE.
19. Tsutsui, S., Yamamura, M., & Higuchi, T. (1999). Multi-parent recombination with simplex crossover in real coded genetic algorithms. In *Proceedings of the 1st Annual Conference on Genetic and Evolutionary Computation-Volume 1* (pp. 657–664). Morgan Kaufmann Publishers Inc.
20. Bck, T., Fogel, D. B., & Michalewicz, Z. (Eds.). (2000). *Evolutionary computation 1: Basic algorithms and operators* (Vol. 1). CRC press.
21. Oliver, I. M., Smith, D., & Holland, J. R. (1987). Study of permutation crossover operators on the travelling salesman problem. In *Proceedings of the Second International Conference on Genetic Algorithms and their Applications*, July 28–31, 1987 at the Massachusetts Institute of Technology, Cambridge, MA. Hillsdale, NJ: L. Erlhaum Associates.
22. Davis, L. (1985). Applying adaptive algorithms to epistatic domains. In *IJCAI* (Vol. 85, pp. 162–164).
23. Whitley, D., Timothy, S., & Daniel, S. Schedule optimization using genetic algorithms. In D. Lawrence (Ed.) 351–357.
24. Grefenstette, J., Gopal, R., Rosmaita, B., & Van Gucht, D. (1985). Genetic algorithms for the traveling salesman problem. In *Proceedings of the first International Conference on Genetic Algorithms and their Applications* (pp. 160–168).
25. Louis, S. J., & Rawlins, G. J. (1991). Designer genetic algorithms: Genetic algorithms in structure design. In *ICGA* (pp. 53–60).
26. Eshelman, L. J., Caruana, R. A., & Schaffer, J. D. (1989). Biases in the crossover landscape. In *Proceedings of the Third International Conference on Genetic Algorithms* (pp. 10–19). Morgan Kaufmann Publishers Inc.
27. Deep, K., & Thakur, M. (2007). A new mutation operator for real coded genetic algorithms. *Applied Mathematics and Computation, 193*(1), 211–230.
28. Srinivas, M., & Patnaik, L. M. (1994). Adaptive probabilities of crossover and mutation in genetic algorithms. *IEEE Transactions on Systems, Man, and Cybernetics, 24*(4), 656–667.
29. Neubauer, A. (1997). A theoretical analysis of the non-uniform mutation operator for the modified genetic algorithm. In *IEEE International Conference on Evolutionary Computation* (pp. 93–96). IEEE.
30. Hinterding, R. (1995). Gaussian mutation and self-adaption for numeric genetic algorithms. In *IEEE International Conference on Evolutionary Computation* (Vol. 1, p. 384). IEEE.
31. Tsutsui, S., & Fujimoto, Y. (1993). Forking genetic algorithm with blocking and shrinking modes (fGA). In *ICGA* (pp. 206–215).
32. Oosthuizen, G. D. (1987). Supergran: A connectionist approach to learning, integrating genetic algorithms and graph induction. In *Proceedings of the second International Conference on Genetic Algorithms and their Applications*, July 28–31, 1987 at the Massachusetts Institute of Technology, Cambridge, MA. Hillsdale, NJ: L. Erlhaum Associates.

33. Mauldin, M. L. (1984). Maintaining diversity in genetic search. In *AAAI* (pp. 247–250).
34. Ankenbrandt, C. A. (1991). An extension to the theory of convergence and a proof of the time complexity of genetic algorithms. In *Foundations of genetic algorithms* (Vol. 1, pp. 53–68). Elsevier.
35. Ahn, C. W., & Ramakrishna, R. S. (2003). Elitism-based compact genetic algorithms. *IEEE Transactions on Evolutionary Computation, 7*(4), 367–385.
36. Mirjalili, S., Gandomi, A. H., Mirjalili, S. Z., Saremi, S., Faris, H., & Mirjalili, S. M. (2017). Salp swarm algorithm: A bio-inspired optimizer for engineering design problems. *Advances in Engineering Software, 114*, 163–191.

Chapter 5
Biogeography-Based Optimisation

5.1 Introduction

Biogeography-Based Optimisation (BBO) [1] is one of the recent evolutionary algorithms with successful application in a diverse field of studies. Similarly to other evolutionary algorithms, BBO has been equipped with crossover and mutations. The main difference between this algorithm and GA is the use of two operators to perform crossover and exploitation. The concepts of mutation is also similar, in which small changes occur in variables of solutions. However, each solution in BBO faces different mutation rates depending on its fitness, which makes it different from the GA algorithm. In this chapter, the inspiration and mathematical equations of the BBO algorithm are first given. A set of problems is then solved with this algorithm to observe and analyse its performance.

5.2 Inspiration

The field of biogeography (or zoogeography) studies the geographical distribution of biological organisms. The evolution of biological organisms can be investigated over time and/or space. In biogeography, the spatial variants of organisms is investigated over time. The main inspiration of the BBO algorithm comes from this field. The BBO algorithm simulates the immigration, emigration, and mutation of different species [2]. The main motivation of developing BBO was the fact that nature finds an optimal balance between preys and predators in an ecosystem (limited in a geographical location) by reducing or increasing the number of organism of species using removal (immigration and/or emigration) or adaptation (mutation).

The term "island" [3] is used in the field of biogeography to refer the isolated areas where different species. In the BBO algorithm, an "island" is called habitat and the species who live in it are called habitant. Each geographical region is different from others in terms of vegetation, rainfall, temperature, humidity, soil type, etc.

© Springer International Publishing AG, part of Springer Nature 2019
S. Mirjalili, *Evolutionary Algorithms and Neural Networks*, Studies
in Computational Intelligence 780, https://doi.org/10.1007/978-3-319-93025-1_5

This set of features define Habitat Suitability Index (HSI), which indicates how well the habitat is for habitants.

5.3 Mathematical Model

The mathematical models in the field of biogeography simulates how species arise, migrate, and become extinct. These three concepts are modelled in the BBO algorithm. To solve optimisation problems using BBO, a solution is equivalent to a habitat and habitants represent the variables. Since the only metric to evaluate solutions is the objective value, the HSI is defined by the objective value of the solution. This is shown in Fig. 5.1.

In nature, when a habitat becomes competitive, species have to adapt using mutation or migrate to avoid extinction. Therefore, there is always a balance between the number of specie and HSI of a habitat. The balance between predators and preys is the results of this phenomenon as well.

The BBO algorithm manipulates the solutions with the following rules [4]:

- Habitats with high HSI face high emigration rates
- Habitats with low HSI face low emigration rates
- Habitants in high HSI tend to emigrate to habitants with low HSI
- There are random changes in the habitants no matter if they are in a high-HSI or low-HSI habitats

In GA, best solutions crossover which increases the probability of population improvement. In the BBO algorithm, the average fitness of population increases since 'bad' variables tend to be replaced by 'good' ones. The mutation is also a good operator to maintain the diversity of solutions.

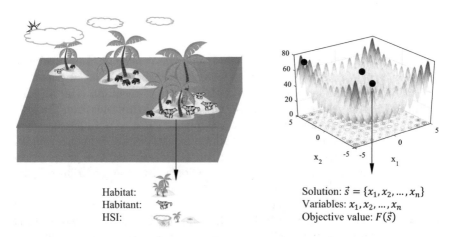

Fig. 5.1 (left) Habitat, habitant, and HST (right) Solution, variable, and objective value in the BBO algorithm

In the BBO algorithm, a population of habitats is first created with randomly initializing the habitants. Each habitat is n dimensional where n is the number of variables for the optimisation problem. To show different exploratory and exploitative behaviour for each solution in BBO, each habitat is assigned with a randomly generated immigration (λ_k), emigration (μ_k), and mutation rate (m_n).

Immigration and emigration rates are calculated as follows:

$$\lambda_k = I \frac{1-n}{N} \tag{5.1}$$

$$\mu_k = \frac{E * n}{N} \tag{5.2}$$

where N is the maximum number of habitants in the habitat, n is the current number of habitants, I shows the maximum immigration rate, and E indicates the maximum emigration rate.

Figure 5.2 shows how each of these rates changes when the number of habitants increases. It can be seen that the immigration rate decreases proportional to the number of habitants. By contrast, the emigration rate increases proportional to the number of iterations. These two equations define how a habitat sends or receives habitants. Note that the maximum number of habitants is increased proportional to HSI. The higher HSI, the more habitants can 'survive' in the habitat.

The mutation rate of a habitat is also defined with the following equation:

$$m_n = M \left(1 - \frac{p_n}{p_{max}} \right) \tag{5.3}$$

where M shows the initial value for the mutation, p_n is the mutation probability in nth habitat, and p_{max} is calculated as follows:

$$p_{max} = argmax(p_n), \quad n = 1, 2, \ldots, N \tag{5.4}$$

Fig. 5.2 Immigration (λ_k) and emigration rates (μ_k) when $N = 50$, $I = 2$, and $E = 1$

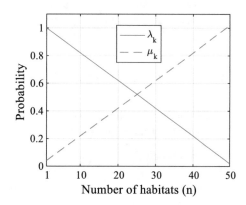

The equation for the mutation shows that this parameter does not change based on the number of habitats or iterations. This is because random changes in the solutions occur all the times regardless of their fitness of habitats. Each habitat has a different mutation rate, which is a good mechanism to improve the diversity of solutions in the BBO algorithm.

An important mechanism in evolutionary algorithm is elitism, in which an algorithm stores the best solution found during the optimisation process since the are likely to be degraded when combining with other solutions. The BBO algorithm also maintains a percentage of best habitats in each iteration.

In the BBO algorithm, the optimisation process starts with a population of habitats with random habitants. The algorithm calculates the HSI, update controlling parameters (λ_k, μ_k, and m_n.), manipulate habitants, and update the elites repeatedly until an end condition is met. At the end of optimisation, the best elite is returned as the best estimation of the global optimum.

5.4 Solving Engineering Design Problems with BBO

To investigate the performance of the BBO algorithm, it is applied to several engineering design problems. BBO was already employed to solve benchmark problems in the first paper on this algorithm, so the engineering problems provide different test beds. The engineering design problems are as follows:

- Three-bar truss design problem
- I-beam design problem
- Welded bean design problem
- Cantilever beam design problem
- Tension/compression spring design problem
- Pressure vessel design problem
- Gear train design problem

The following sub-sections first present the details of these problems. The results of BBO are then given and compared to some of the existing works in the literature.

5.4.1 Three-Bar Truss Design Problem

As the name implies, this problem deals with designing a truss with three bars. The mathematical formulation of this problem is as follows:

$$\text{Consider} : \overrightarrow{x} = [x_1, x_2] = [A_1, A_2] \tag{5.5}$$

$$\text{Minimise} : f(\overrightarrow{x}) = \left(2\sqrt{2}x_1 + x_2\right) \times l \tag{5.6}$$

Fig. 5.3 Three-bar truss design problem

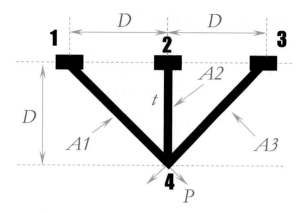

Table 5.1 The results of BBO and other existing algorithms when solving the three-bar truss design problem.

Algorithm	Optimal values for variables		Optimal weight
	x1	x2	
BBO	0.05448316	0.42767526	263.8958517
DEDS [5]	0.78867513	0.40824828	263.8958434
PSO-DE [6]	0.7886751	0.4082482	263.8958433
MBA [7]	0.788565	0.4085597	263.8958522
Ray and Sain [8]	0.795	0.395	264.3
Tsa [9]	0.788	0.408	263.68
CS [10]	0.78867	0.40902	263.9716

$$\text{Subject to} : g_1(\overrightarrow{x}) = \frac{sqrt2x_1 + x_2}{sqrt2x_1^2 + 2x_1x_2} P - \sigma \leq 0 \qquad (5.7)$$

$$g_2(\overrightarrow{x}) = \frac{x_2}{sqrt2x_1^2 + 2x_1x_2} P - \sigma \leq 0 \qquad (5.8)$$

$$g_3(\overrightarrow{x}) = \frac{1}{\sqrt{2}x_2 + x_1} P - \sigma \leq 0 \qquad (5.9)$$

$$\text{Varibles range}: 0 \leq x_1, x_2 \leq 1 \qquad (5.10)$$

where $l = 100\,cm$, $P = 2\,KN/cm^2$, and $\sigma = 2\,KN/cm^2$

The shape of the truss and structural parameters are visualized in Fig. 5.3. It can be seen that there are two parameters to be optimised. The objective is to for the optimal values for both parameters to minimise the weight of the truss subject to three constrains: g_1, g_2, and g_3.

This problem is solved by the BBO algorithm and the results are compared with six existing solutions in Table 5.1. This table shows that the BBO algorithm finds a competitive optimal solution for this problem.

5.4.2 I-Beam Design Problem

The I-beam design is a structural design problem, in which the main objective is to minimise the vertical deflection of an I-beam. The mathematical formulation and shape of this beam are given below:

$$\text{Consider: } \overrightarrow{x} = [x_1, x_2, x_3, x_4, x_5] = \left[b, h, t_w, t_f \right] \tag{5.11}$$

$$\text{Minimise: } f(\overrightarrow{x}) = \frac{5000}{\frac{t_w(h-2t_f)^3}{12} + \frac{bt_f^3}{6} + 2bt_f \left(\frac{h-t_f}{2} \right)^2} \tag{5.12}$$

$$\text{Subject to: } g(\overrightarrow{x}) = 2bt_w + t_w(h - 2t_f) \leq 0 \tag{5.13}$$

$$\text{Variables range: } 10 \leq x_1 \leq 50 \tag{5.14}$$

$$10 \leq x_2 \leq 80 \tag{5.15}$$

$$0.9 \leq x_3 \leq 5 \tag{5.16}$$

$$0.9 \leq x_4 \leq 5 \tag{5.17}$$

It can be observed in the equations and Fig. 5.4 that there are four structural parameters to optimise:

- Length (L)
- Width (b)
- Thickness of the vertical bar (T_w)
- Thickness of the horizontal bar (T_f)

The results of the BBO algorithm on this problem are presented and compared with four existing solutions in Table 5.2. This table shows that the BBO algorithm provides the best solution for this problem. The discrepancy of the results is significant that shows the high performance of BBO on this problem.

Fig. 5.4 The structural parameter of the I-beam design problem

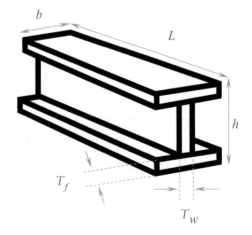

Table 5.2 The results of BBO and other existing algorithms when solving the I-beam design problem.

Algorithm	Optimal values for variables				Optimum vertical deflection
	b	h	t_w	t_f	
BBO	50	80	1.76	5	0.0066259
ARSM [11]	37.05	80	1.71	2.31	0.0157
IARSM [11]	48.42	79.99	0.9	2.4	0.131
CS [10]	50	80	0.9	2.321675	0.0130747
SOS [12]	50	80	0.9	2.32179	0.0130741

5.4.3 Welded Beam Design Problem

The mathematical formulations of this problem are as follows:

$$\text{Consider: } \vec{x} = [x_1, x_2, x_3, x_4] = [h, l, t, b] \qquad (5.18)$$

$$\text{Minimise:} f(\vec{x}) = 1.10471x_1^2 x_2 + 0.04811 x_3 x_4 (14.0 + x_2) \qquad (5.19)$$

$$\text{Subject to: } g_1(\vec{x}) = \qquad (5.20)$$

$$\tau(\vec{x}) - \tau_{max} \le 0 \qquad (5.21)$$

$$g_2(\vec{x}) = \qquad (5.22)$$

$$\sigma(\vec{x}) - \sigma_{max} \le 0 \qquad (5.23)$$

$$g_3(\vec{x}) = \qquad (5.24)$$

$$x_1 - x_4 \le 0 \qquad (5.25)$$

$$g_4(\vec{x}) = \qquad (5.26)$$

$$1.10471x_1^2 x_2 + 0.04811 x_3 x_4 (14.0 + x_2) - 5 \le 0 \qquad (5.27)$$

$$g_5(\vec{x}) = \qquad (5.28)$$

$$0.125 - x_1 \le 0 \qquad (5.29)$$

$$g_6(\vec{x}) = \qquad (5.30)$$

$$\delta(\vec{x}) - \delta max \le 0 \qquad (5.31)$$

$$g_7(\vec{x}) = \qquad (5.32)$$

$$P(\vec{x}) - P_c(\vec{x}) \le 0 \qquad (5.33)$$

$$\text{Variables range: } 0.1 \le x_1 \le 2 \tag{5.34}$$

$$0.1 \le x_2 \le 10 \tag{5.35}$$

$$0.1 \le x_3 \le 10 \tag{5.36}$$

$$0.1 \le x_4 \le 2 \tag{5.37}$$

$$\text{where: } \tau(\overrightarrow{x}) = \sqrt{(\tau')^2 + 2\tau'\tau''\frac{x_2}{2R} + (\tau'')^2} \tag{5.38}$$

$$\tau' = \frac{P}{\sqrt{2}x_1x_2}, \tau'' = \frac{MR}{J}, M = P\left(L + \frac{x_2}{2}\right) \tag{5.39}$$

$$R = \sqrt{\frac{x_2^2}{4} + \left(\frac{x_1 + x_3}{2}\right)^2} \tag{5.40}$$

$$J = 2\{\sqrt{2}x_1x_2\left[\frac{x_2^2}{4} + \left(\frac{x_1 + x_3}{2}\right)^2\right]\} \tag{5.41}$$

$$\sigma(\overrightarrow{x}) = \frac{6PL}{x_4x_3^2}, \delta(\overrightarrow{x}) = \frac{6PL^3}{Ex_3^2x_4} \tag{5.42}$$

$$P_c(\overrightarrow{x}) = \frac{4.01E\sqrt{\frac{x_3^2x_4^6}{36}}}{L^2}\left(1 - \frac{x_3}{2L}\sqrt{\frac{E}{4G}}\right) \tag{5.43}$$

The formulation shows that the welded bean design problem has four parameter and the objective is to minimise the fabrication cost. The four structural parameters in this problem are visualized in Fig. 5.5. The BBO algorithm is applied to this problem, and the results are given in Table 5.3. This table shows that there are many existing solutions to this problem, and the BBO algorithm finds one of the best solutions. The CPSO algorithm shows very competitive results.

Fig. 5.5 The structural parameter of the welded beam design problem

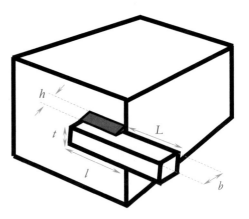

Table 5.3 The results of BBO and other existing algorithms when solving the welded beam design problem

Algorithm	Optimal values for variables				Optimal cost
	h	l	t	b	
BBO	0.205548	3.474442	9.036521	0.205735	1.72512
GSA	0.182129	3.856979	10	0.202376	1.87995
CPSO [13]	0.202369	3.544214	9.04821	0.205723	1.73148
GA [14]	0.1829	4.0483	9.3666	0.2059	1.8242
GA [15]	0.2489	6.173	8.1789	0.2533	2.43312
Coello [16]	0.2088	3.4205	8.9975	0.21	1.74831
Coello and Montes [17]	0.205986	3.471328	9.020224	0.20648	1.72822
Siddall [18]	0.2444	6.2189	8.2915	0.2444	2.38154
Ragsdell [19]	0.2455	6.196	8.273	0.2455	2.38594
Random [19]	0.4575	4.7313	5.0853	0.66	4.11856
Simplex [19]	0.2792	5.6256	7.7512	0.2796	2.53073
David [19]	0.2434	6.2552	8.2915	0.2444	2.38411
APPROX [19]	0.2444	6.2189	8.2915	0.2444	2.38154

5.4.4 Cantilever Beam Design Problem

A cantilever beam in this problem is made of five hollow rectangular blocks as shown in Fig. 5.6. The shape of each block is defined by one parameter, so there are four parameters to optimise in this problem. The details of the objective function, constraints, and range of variables for this problem are as follows:

$$\text{Consider} : \overrightarrow{x} = [x_1, x_2, x_3, x_4, x_5] \qquad (5.44)$$

$$\text{Minimise} : f(\overrightarrow{x}) = 0.6224(x_1 + x_2 + x_3 + x_4 + x_5) \qquad (5.45)$$

$$\text{Subject to} : g(\overrightarrow{x}) = \frac{61}{x_1^3} + \frac{27}{x_2^3} + \frac{19}{x_3^3} + \frac{7}{x_4^3} + \frac{1}{x_5^3} - 1 \leq 0 \qquad (5.46)$$

$$\text{Variables range} : 0.01 \leq x_1, x_2, x_3, x_4, x_5 \leq 100 \qquad (5.47)$$

This problem is solved with the BBP algorithm as well. A comparison is given in Table 5.4, in which the results are compared with those of five other stochastic algorithms in the literature.

Table 5.4 shows that three algorithms (including BBO) find the same best optimal solutions for this problem. These results show that the BBO is very competitive in this problem as well.

Fig. 5.6 The structural
parameter of the cantilever
beam design problem

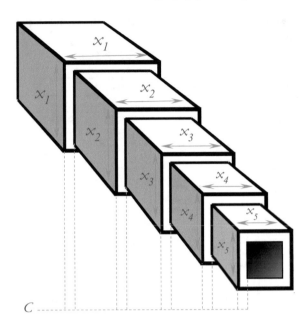

Table 5.4 The results of BBO and other existing algorithms when solving the cantilever beam
design problem

Algorithm	Optimal values for variables					Optimum weight
	x1	x2	x3	x4	x5	
BBO	6.02565	5.30573	4.48081	3.50299	2.15864	1.33996
SOS [12]	6.01878	5.30344	4.49587	3.49896	2.15564	1.33996
MMA [20]	6.01	5.3	4.49	3.49	2.15	1.34
GCA_I [20]	6.01	5.3	4.49	3.49	2.15	1.34
GCA_II [20]	6.01	5.3	4.49	3.49	2.15	1.34
CS [10]	6.0089	5.3049	4.5023	3.5077	2.1504	1.33999

5.4.5 Tension/compression Spring Design

Tension/compression spring design is another problem solved in this chapter. As can
be seen in the following equations, the objective is to minimise the fabrication cost
given three variables: wire diameter (d), mean coil diameter (D), and the number
of active coils (N). There are also four constraints that should not be violated by an
optimisation algorithm.

Fig. 5.7 The structural
parameter of the
tension/compression spring
design problem

$$\text{Consider}: \overrightarrow{x} = [x_1, x_2, x_3] = [d, D, N] \quad (5.48)$$

$$\text{Minimise}: f(\overrightarrow{x}) = (x_3 + 2)x_2 x_1^2 \quad (5.49)$$

$$\text{Subject to}: g_1(\overrightarrow{x}) = 1 - \frac{x_2^3 x_3}{71,785 x_1^4} \leq 0 \quad (5.50)$$

$$g_2(\overrightarrow{x}) = \frac{4x_2^2 - x_1 x_2}{12,566(x_2 x_1^3 - x_1^4)} + \frac{1}{5108 x_1^2} \leq 0 \quad (5.51)$$

$$g_3(\overrightarrow{x}) = 1 - \frac{140.45 x_1}{x_2^2 x_3} \leq 0 \quad (5.52)$$

$$g_4(\overrightarrow{x}) = \frac{x_1 + x_2}{1.5} - 1 \leq 0 \quad (5.53)$$

$$\text{Variables range}: 0.05 \leq x_1 \leq 2.00 \quad (5.54)$$

$$0.15 \leq x_2 \leq 1.3 - \quad (5.55)$$

$$2.00 \leq x_3 \leq 15.0 \quad (5.56)$$

The shape of the tension/compression spring and its structural parameters are presented in Fig. 5.7.

The experimental results of this subsection are given in Table 5.5. It can be seen that BBO is compared with nine algorithms including conventional mathematical and stochastic optimisation techniques.

The results of BBO on this problem are very competitive. This algorithm does not find the best solutions, yet the discrepancy of the objective value is not significant.

5.4.6 Pressure Vessel Design

This problem has four parameters and four constraints. The objective is to minimise the fabrication cost subject to not violating the four constraints. The pressure design problem is formulated as follows:

Table 5.5 The results of BBO and other existing algorithms when solving the tension/compression spring design problem

Algorithm	Optimum variables			Optimum weight
	d	D	N	
BBO	0.051045	0.341417	12.246486	0.0126737
GSA [21]	0.050276	0.32368	13.52541	0.0127022
PSO [13]	0.051728	0.357644	11.244543	0.0126747
ES [22]	0.051989	0.363965	10.890522	0.012681
GA (Coello) [23]	0.05148	0.351661	11.632201	0.0127048
RO [24]	0.05137	0.349096	11.76279	0.0126788
Improved HS [25]	0.051154	0.349871	12.076432	0.0126706
DE 8 [26]	0.051609	0.354714	11.410831	0.0126702
Mathematical optimisation [27]	0.053396	0.39918	9.1854	0.0127303
Constraint correction [28]	0.05	0.3159	14.25	0.0128334

$$\text{Consider}: \vec{x} = [x_1, x_2, x_3, x_4] = [T_s, T_h, R, L] \quad (5.57)$$
$$\text{Minimise}: f(\vec{x}) = 0.6224x_1x_3x_4 + 1.7781x_2x_3^2 + 3.1661x_1^2x_4 + 19.84x_1^2x_3 \quad (5.58)$$
$$\text{Subject to}: g_1(\vec{x}) = -x_1 + 0.0193x_3 \leq 0 \quad (5.59)$$
$$g_2(\vec{x}) == x_3 + 0.00954x_3 \leq 0 \quad (5.60)$$
$$g_3(\vec{x}) = -\pi x_3^2 x_4 - \frac{4}{3}\pi x_3^3 + 1296000 \leq 0 \quad (5.61)$$
$$g_4(\vec{x}) = x_4 - 240 \leq 0 \quad (5.62)$$
$$\text{Variables range}: 0 \leq x_1 \leq 99 \quad (5.63)$$
$$0 \leq x_2 \leq 99 \quad (5.64)$$
$$10 \leq x_3 \leq 200 \quad (5.65)$$
$$10 \leq x_4 \leq 200 \quad (5.66)$$

The shape of the pressure vessel and the structural parameters are show in Fig. 5.8. It can be seen that the structure parameters are: L, R, T_h, and T_s.

Table 5.6 shows the results of BBO and six other algorithms on this problem, in which the superior performance of BBO is evident.

5.4.7 Gear Train Design Problem

The last problem solved in this chapter is the gear train design problem. This is the only discrete problem, and to find integer values for the four structural parameters, BBO is required to round the solutions to the nearest integer before every function

Fig. 5.8 The structural parameters of the pressure vessel design problem

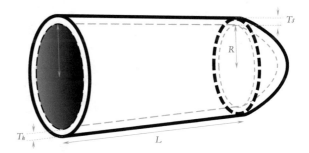

Table 5.6 The results of BBO and other existing algorithms when solving the pressure vessel design problem

Algorithm	Optimum variables				Optimum cost
	T_s	T_h	R	L	
BBO	0.85993	0.425225	44.553437	148.419383	6041.1894
CGSA9 (Sinusoidal) [29]	0.890827	0.439804	45.86797	135.1164	6054.2419
GSA [21]	1.080581	7.897191	55.988659	84.4542025	48807.29
GA (Coello) [23]	0.8125	0.4345	40.3239	200	6288.7445
GA (Deb and Gene) [30]	0.9375	0.5	48.329	112.679	6410.3811
Lagrangian Multiplie (Kannan) [31]	1.125	0.625	58.291	43.69	7198.0428
branch-bound (Sandgren) [32]	1.125	0.625	47.7	117.701	8129.1036

evaluation. Each parameter indicates the number of teeth in a cog system as shows in Fig. 5.9. Also, there is no constraints, so this problem is the only unconstrained problem solve in this chapter.

The problem formulation of the gear train design is shown as follows:

$$\text{Consider} : \vec{x} = [x_1, x_2, x_3, x_4] = [n_A, n_B, n_C, n_D] \qquad (5.67)$$

$$\text{Minimise} : f\left(\vec{x}\right) = \left(\frac{1}{6.931} - \frac{x_3 x_2}{x_1 x_4}\right)^2 \qquad (5.68)$$

$$\text{Subject to} : g\left(\vec{x}\right) = 12 \leq x_1, x_2, x_3, x_4 \leq 60 \qquad (5.69)$$

Fig. 5.9 The structural parameters of the gear train design problem

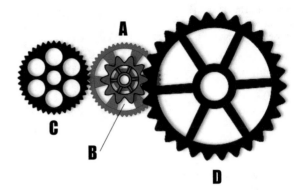

Table 5.7 The results of BBO and other existing algorithms when solving the gear train design problem

Algorithm	Optimal values for variables				f
	n_1	n_2	n_3	n_4	
BBO	49	18	15	43	2.70E-12
ABC [7]	49	16	19	43	2.70E-12
MBA [7]	43	16	19	49	2.70E-12
GA [33]	49	16	19	43	2.701 9e-012
CS [10]	43	16	19	49	2.70E-12
ISA [34]	N/A	N/A	N/A	N/A	2.70E-12
Kannan and Kramer [31]	33	15	13	41	2.146 9e-08

Table 5.7 shows the results of BBO and six comparative algorithms. Most of the algorithms find an equal optimal objective value, which is due to the simplicity fo this problem and the lack of constraints. However, it seems the design found by the BBO algorithm is slightly different since it is the only algorithm that finds the value of 15 for the third parameter.

Overall, the results of chapter showed that the BBO algorithm is able to provide very comparative, often superior performance as compared to a large number of algorithms in the literature. BBO is an evolutionary algorithm equipped with two new evolutionary operator: immigration and emigration. These two operators promote exploitation and local search since they combine solutions. The BBO algorithm also uses several mutations to change habitants abruptly and do a global search. Adaptive parameters of BBO (immigration and emigration rates) balance the contradictory phases of exploration and exploitation. These are the reasons for the high performance of BBO when solving the problems in the chapter. This algorithm will be applied to NNs in the second part of this book to benefit from its advantages.

References

1. Simon, D. (2008). Biogeography-based optimization. *IEEE Transactions on Evolutionary Computation, 12*(6), 702–713.
2. Mirjalili, S., Mirjalili, S. M., & Lewis, A. (2014). Let a biogeography-based optimizer train your multi-layer perceptron. *Information Sciences, 269*, 188–209.
3. MacArthur, R. H., & Wilson, E. O. (2016). *The theory of Island biogeography*. Princeton University Press.
4. Saremi, S., Mirjalili, S., & Lewis, A. (2014). Biogeography-based optimisation with chaos. *Neural Computing and Applications, 25*(5), 1077–1097.
5. Zhang, M., Luo, W., & Wang, X. (2008). Differential evolution with dynamic stochastic selection for constrained optimization. *Information Sciences, 178*(15), 3043–3074.
6. Liu, H., Cai, Z., & Wang, Y. (2010). Hybridizing particle swarm optimization with differential evolution for constrained numerical and engineering optimization. *Applied Soft Computing, 10*(2), 629–640.
7. Sadollah, A., Bahreininejad, A., Eskandar, H., & Hamdi, M. (2013). Mine blast algorithm: A new population based algorithm for solving constrained engineering optimization problems. *Applied Soft Computing, 13*(5), 2592–2612.
8. Ray, T., & Saini, P. (2001). Engineering design optimization using a swarm with an intelligent information sharing among individuals. *Engineering Optimization, 33*(6), 735–748.
9. Tsai, J. F. (2005). Global optimization of nonlinear fractional programming problems in engineering design. *Engineering Optimization, 37*(4), 399–409.
10. Gandomi, A. H., Yang, X. S., & Alavi, A. H. (2013). Cuckoo search algorithm: A metaheuristic approach to solve structural optimization problems. *Engineering with Computers, 29*(1), 17–35.
11. Wang, G. G. (2003). Adaptive response surface method using inherited Latin hypercube design points. *Journal of Mechanical Design, 125*(2), 210–220.
12. Cheng, M. Y., & Prayogo, D. (2014). Symbiotic organisms search: A new metaheuristic optimization algorithm. *Computers & Structures, 139*, 98–112.
13. He, Q., & Wang, L. (2007). An effective co-evolutionary particle swarm optimization for constrained engineering design problems. *Engineering Applications of Artificial Intelligence, 20*(1), 89–99.
14. Coello Coello, C. A. (2000). Constraint-handling using an evolutionary multiobjective optimization technique. *Civil Engineering Systems, 17*(4), 319–346.
15. Deb, K. (2000). An efficient constraint handling method for genetic algorithms. *Computer Methods in Applied Mechanics and Engineering, 186*(2–4), 311–338.
16. Coello, C. A. C. (2002). Theoretical and numerical constraint-handling techniques used with evolutionary algorithms: A survey of the state of the art. *Computer Methods in Applied Mechanics and Engineering, 191*(11–12), 1245–1287.
17. Coello, C. A. C., & Montes, E. M. (2002). Constraint-handling in genetic algorithms through the use of dominance-based tournament selection. *Advanced Engineering Informatics, 16*(3), 193–203.
18. Siddall, J. N. (1972). *Analytical decision-making in engineering design*. Prentice Hall.
19. Ragsdell, K. M., & Phillips, D. T. (1976). Optimal design of a class of welded structures using geometric programming. *Journal of Engineering for Industry, 98*(3), 1021–1025.
20. Chickermane, H. E. M. I. A. N. T., & Gea, H. C. (1996). Structural optimization using a new local approximation method. *International Journal for Numerical Methods in Engineering, 39*(5), 829–846.
21. Mirjalili, S. (2015). Moth-flame optimization algorithm: A novel nature-inspired heuristic paradigm. *Knowledge-Based Systems, 89*, 228–249.
22. Mezura-Montes, E., & Coello, C. A. C. (2008). An empirical study about the usefulness of evolution strategies to solve constrained optimization problems. *International Journal of General Systems, 37*(4), 443–473.
23. Coello, C. A. C. (2000). Use of a self-adaptive penalty approach for engineering optimization problems. *Computers in Industry, 41*(2), 113–127.

24. Kaveh, A., & Khayatazad, M. (2012). A new meta-heuristic method: Ray optimization. *Computers & Structures, 112*, 283–294.

25. Mahdavi, M., Fesanghary, M., & Damangir, E. (2007). An improved harmony search algorithm for solving optimization problems. *Applied Mathematics and Computation, 188*(2), 1567–1579.

26. Li, L. J., Huang, Z. B., Liu, F., & Wu, Q. H. (2007). A heuristic particle swarm optimizer for optimization of pin connected structures. *Computers & Structures, 85*(7–8), 340–349.

27. Belegundu, A. D., & Arora, J. S. (1985). A study of mathematical programming methods for structural optimization. Part I: Theory. *International Journal for Numerical Methods in Engineering, 21*(9), 1583–1599.

28. Arora, J. (2004). *Introduction to optimum design*. Elsevier.

29. Mirjalili, S., & Gandomi, A. H. (2017). Chaotic gravitational constants for the gravitational search algorithm. *Applied Soft Computing, 53*, 407–419.

30. Deb, K. (1997). GeneAS: A robust optimal design technique for mechanical component design. In *Evolutionary algorithms in engineering applications* (pp. 497–514). Heidelberg: Springer.

31. Kannan, B. K., & Kramer, S. N. (1994). An augmented Lagrange multiplier based method for mixed integer discrete continuous optimization and its applications to mechanical design. *Journal of Mechanical Design, 116*(2), 405–411.

32. Sandgren, E. (1990). Nonlinear integer and discrete programming in mechanical design optimization. *Journal of Mechanical Design, 112*(2), 223–229.

33. Deb, K., & Goyal, M. (1996). A combined genetic adaptive search (GeneAS) for engineering design. *Computer Science and Informatics, 26*, 30–45.

34. Gandomi, A. H. (2014). Interior search algorithm (ISA): A novel approach for global optimization. *ISA Transactions, 53*(4), 1168–1183.

Part II
Evolutionary Neural Networks

In the second part of the book, the algorithms presented in the first part are employed to design evolutionary training algorithms for different types of NNs including:

- Feedforward neural networks
- Multi-layer perceptrons
- Radial basis function networks
- Deep neural network

Each chapter starts with a brief literature review to show the state of the art in each class of NNs. The method of designing an evolutionary training algorithm is then discussed in detail. There are several experiments to investigate the efficiency of the evolutionary training algorithms. It is tried to use test and real-world datasets with different levels of difficulty to compare the algorithms from different perspectives.

Chapter 6
Evolutionary Feedforward Neural Networks

6.1 Introduction

Feedforward Neural Networks (FNN) have been of the most popular NNs with a wide range of applications. The process of finding optimal values for controlling parameters of a NN is called training and can be considered as an optimisation problem. In this optimisation problem, the ultimate goal is to find optimal values for reduce the errors of the NNs or improves its accuracy. The conventional training algorithm for FNNs is called Back Propagation (BP). This trainer is gradient-based, which uses a gradient descent operator to find the closet optimal solution from a given random initial point. However, the BP algorithm suffers from local optima stagnation due to the use of gradient [1]. In this chapter several evolutionary algorithm including the ones presented in Part I of this book are employed to train FNN as reliable alternative to the BP algorithm.

6.2 Feedforward Neural Networks

FNNs are those NNs with only one-way and one-directional connection between their neurons. In this type of NNs, neurons are arranged in different parallel layers [2]. The first layer is always called the input layer, whereas the last later is called the output layers. Other layers between the input and output layers are called hidden layer. A FNN with one hidden layer is called FNN as illustrated in Fig. 6.1.

After providing the inputs, weights, and biases, the output of FNNs are calculated throughout the following steps:

1. The weighted sums of inputs are first calculated by Eq. 6.1.

$$s_j = \sum_{i=1}^{n} \left(W_{ij} X_i \right) - \theta_j, \quad j = 1, 2, ..., h \qquad (6.1)$$

© Springer International Publishing AG, part of Springer Nature 2019
S. Mirjalili, *Evolutionary Algorithms and Neural Networks*, Studies
in Computational Intelligence 780, https://doi.org/10.1007/978-3-319-93025-1_6

Fig. 6.1 FNN with one
hidden layer

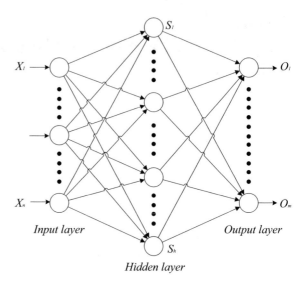

S_1

X_1 → ○ → O_1

X_n → O_m

Input layer   *Output layer*

S_h

Hidden layer

where n is the number of the input nodes, W_{ij} shows the connection weight from
the *ith* node in the input layer to the *jth* node in the hidden layer, θ_j is the bias
(threshold) of the *jth* hidden node, and X_i indicates the *ith* input.

2. The output of each hidden node is calculated as follows:

$$S_j = sigmoid\,(s)j) = \frac{1}{1 + e^{-s_j}}, \qquad j = 1, 2, ...m \qquad (6.2)$$

3. The final outputs are defined based on the calculated outputs of the hidden nodes
as follows:

$$o_k = \sum_{j=1}^{h} \left(w_{jk}S_j\right) - \theta'_k, \qquad k = 1, 2, ..., m \qquad (6.3)$$

$$O_k = sigmoid\,(o_k) = \frac{1}{1 + e^{-o_k}}, \qquad k = 1, 2, ..., m \qquad (6.4)$$

where w_{jk} is the connection weight from the jth hidden node to the *kth* output
node, and $'_k$ is the bias (threshold) of the *kth* output node.

As may be seen in the Eqs. 6.1–6.4, the weights and biases are responsible for
defining the final output of FNNs from given inputs. Finding proper values for weights
and biases in order to achieve a desirable relation between the inputs and outputs is
the exact definition of training FNNs. In the next sections, the GWO algorithm is
employed as a trainer for MLPs.

6.3   Designing Evolutionary FNNs

The first and most important step in training an FNN using meta-heuristics is the
problem representation [3]. In other words, the problem of training FNNs should
be formulated in a way that is suitable for meta-heuristics. As mentioned in the
introduction, the most important variables in training an FNN are weights and biases.
A trainer should find a set of values for weights and biases that provide the highest
classification/approximation/prediction accuracy.

Therefore, the variables here are the weights and biases. Since most of EA accept
the variables in the form of a vector, the variables of an FNN are presented for this
algorithm as follows:

$$\{\overrightarrow{W}, \overrightarrow{\theta}\} = \{W_{1,1}, W_{1,2}, ..., W_{n,h}, \theta_1, \theta_2, ..., \theta_h\} \tag{6.5}$$

where n is the number of the input nodes, W_{ij} shows the connection weight from the
ith node in the input layer to the *jth* node in the hidden layer, θ_j is the bias (threshold)
of the *jth* hidden node.

After defining the variables, we need to define the objective function for EAs. As
mentioned above, the objective in training an FNN is to reach the highest classification, approximation, or predication accuracy for both training and testing samples.
A common metric for the evaluation of an FNN is the Mean Square Error (MSE). In
this metric, a given set of training samples is applied to the FNN and the following
equation calculates the difference between the desirable output and the value that is
obtained from the FNN:

$$MSE = \sum_{i=1}^{m} \left(o_i^k - d_i^k\right)^2 \tag{6.6}$$

where m is the number of outputs, d_i^k is the desired output of the *ith* input unit when
the *kth* training sample is used, and o_i^k is the actual output of the *ith* input unit when
the *kth* training sample appears in the input.

Obviously, an FNN should adapt itself to the whole set of training samples in
order to be effective. Therefore, the performance of an FNN is evaluated based on
the average of MSE over all the training samples as follows:

$$\overline{MSE} = \sum_{k=1}^{s} \frac{\sum_{i=1}^{m}(o_i^k - d_i^k)^2}{s} \tag{6.7}$$

where s is the number of training samples, m is the number of outputs, d_i^k is the
desired output of the ith input unit when the *kth* training sample is used, and o_i^k is the
actual output of the *ith* input unit when the *kth* training sample appears in the input.

Fig. 6.2 Evolutionary
algorithms provide FNN
with weights/biases and
received average MSE for all
training samples

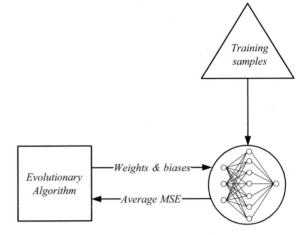

After all, the problem of training an FNN can be formulated with the variable and average MSE for EAs as follows:

$$Minimise: \quad F(\vec{x} = \overline{MSE}) \tag{6.8}$$

Figure 6.2 shows the overall process of training an FNN using EA [11]. It may be seen that the GWO algorithm provides FNN with weights/biases and receives average MSE for all training samples. The EA algorithm iteratively changes the weights and biases to minimise average MSE of all training samples.

6.4   Results

In this section several evolutionary trainers using Grey Wolf Optimiser (GWO) [4], PSO [5], GA [6], ACO [7], ES [8], and Population-based Incremental Learning (PBIL) [9] are applied to standard classification datasets obtained from the University of California at Irvine (UCI) Machine Learning Repository [10]: XOR, balloon, iris, breast cancer, and heart. Other assumptions for the algorithms are presented in Table 6.1.

Table 6.2 the specifications of the datasets. It can be observed in table that the easiest dataset is XOR with 8 training/test samples, 3 attributes, and two classes. The balloon dataset is more difficult that the XOR and has 4 attributes, 16 training samples, 16 test samples, and 2 classes. The Iris dataset has 4 attributes, 150 training/test samples, and three classes. In addition, the breast cancer dataset has 9 attributes, 599 training samples, 100 test samples, and 2 classes. Eventually, the heart dataset includes 22 attributes, 80 training samples, 187 test samples, and 2 classes. These classification datasets were deliberately chosen with different training/test samples and levels of difficulty to test the performance of the evolutionary trainers effectively.

Table 6.1 The initial parameters of algorithms

Algorithm	Parameter	Value
GWO	α	linearly decreased from 2 to 0
	Population size	50 for XOR and Balloon, 200 for the rest
	Maximum number of generations	250
PSO	Topology	connected
	Cognitive constant (c_1)	1
	Social constant (c_2)	1
	Inertia constant (w)	0.3
	Population size	50 for XOR and Balloon, 200 for the rest
	Maximum number of iterations	250
GA	Type	Real coded
	Selection	Roulette wheel
	Crossover	Single point (probability = 1)
	Mutation	Uniform (probability = 0.01)
	Population size	50 for XOR and Balloon, 200 for the rest
	Maximum number of generations	250
ACO	Initial pheromone (τ_0)	1e-06
	Pheromone update constant (Q)	20
	Pheromone constant (q_0)	1
	Global pheromone decay rate (p_g)	0.9
	Local pheromone decay rate (p_t)	0.5
	Pheromone sensitivity (α)	1
	Visibility sensitivity (β)	5
	Population size	50 for XOR and Balloon, 200 for the rest
	Maximum number of iterations	250
ES	λ	10
	σ	1
	Population size	50 for XOR and Balloon, 200 for the rest
	Maximum number of generations	250
PBIL	Learning rate	0.05
	Good population member	1
	Bad population member	0
	Elitism parameter	1
	Mutational probability	0.1
	Population size	50 for XOR and Balloon, 200 for the rest
	Maximum number of generations	250

A similar problem representation and objective function is utilised to train FNN with the algorithms in Table 6.1. The datasets are then solved 10 times using each algorithm to generate the results. The statistical results that are presented are average (AVE) and standard deviation (STD) of the obtained best MSEs in the last iteration by the algorithms. Obviously, lower average and standard deviation of MSE in the last

Table 6.2 Classification datasets

Classification datasets	Number of attributes	Number of training samples	Number of test samples	Number of classes
3-bits XOR	3	8	8 as training samples	2
Balloon	4	16	16 as training samples	2
Iris	4	150	150 as training samples	3
Breast cancer	9	599	100	2
Heart	22	80	187	2

iteration indicates the better performance. We conduct the experiment and present the statistical results in the form of (AVE $\pm$ STD) in Tables 6.4, 6.5, 6.6, 6.7 and 6.8. Please note that the best classification rates or test errors obtained by each of the algorithms during 10 runs are reported as another metrics of comparison. Normalization is and essential steps for FNN when solving datasets with attributes in different ranges.

The normalization used in this work is called min-max normalization that is formulated as follows:

$$x' = \frac{(x - a) \times (d - c)}{(b - a)} + c \qquad (6.9)$$

This formula maps x in the interval of $[a, b]$ to $[c, d]$. Another key factor in the experimental setup is the structure of FNNs. This work does not concentrate on finding the optimal number of hidden nodes and consider them equal to $2N + 1$ where N is the number of features (inputs) of the datasets. The structure of each FNN that employed for each dataset is presented in Table 6.3.

As the size of the neural network becomes larger, obviously, the more weights and biases would be involved in the system. Consequently, the training process also becomes more challenging.

Table 6.3 FNN structure for each dataset

Classification datasets	Number of attributes	FNN structure
3-bits XOR	3	3-7-1
Balloon	4	4-9-1
Iris	4	4-9-3
Breast cancer	9	9-19-1
Heart	22	22-45-1

Table 6.4 Experimental results for the XOR dataset

Algorithm	MSE (AVE $\pm$ STD)	Classification rate (%)
GWO-FNN	0.009410 $\pm$ 0.029500	100.00
PSO-FNN	0.084050 $\pm$ 0.035945	37.50
GA-FNN	0.000181 $\pm$ 0.000413	100.00
ACO-FNN	0.180328 $\pm$ 0.025268	62.50
ES-FNN	0.118739 $\pm$ 0.011574	62.50
PBIL-FNN	0.030228 $\pm$ 0.039668	62.50

6.4.1 XOR Dataset

This dataset has 3 inputs, 8 training/test samples, and 1 output. The FNN for this dataset has to return the XOR of the input as the output. The statistical results of GWO, PSO, GA, ACO, ES, and PBIL on this dataset are presented in Table 6.5. Note that the training algorithms are named in the form of Algorithm-FNN in Table 6.4.

The results show that the best average MSE belongs to GA-FNN and GWO-FNN. This shows that these two algorithms have the highest ability to avoid local optima, significantly better than other algorithms. GWO-FNN and GA-FNN also classifies this dataset with 100 percent accuracy. The GA-FNN algorithm is an evolutionary algorithm which has a very high level of exploration. These results show that the GWO-based trainer is able to show very competitive results compare to the GA-FNN algorithm.

Another fact worth noticing here is the poor performance of ACO-FNN on this dataset. This might be due to the fact that this algorithm is more suitable for combinatorial problems.

6.4.2 Balloon Dataset

This dataset has 4 attributes, 18 training/test samples, and 2 classes. The trainers for this dataset have dimensions of 55. The results are provided in Table 6.5.

The first thing that can be observed in the results is the similar classification rate for all algorithms. This behaviour is due to simplicity of this dataset. However, the average and standard deviation of the MSE over 10 runs are different for algorithms. Similarly to the previous dataset, GWO-FNN and GA-FNN show the best local optima avoidance as per the statistical results of MSEs. This again justifies high performance of the GWO algorithm in training FNNs.

Table 6.5 Experimental results for the balloon dataset

Algorithm	MSE (AVE ± STD)	Classification rate (%)
GWO-FNN	9.38e-15 ± 2.81e-14	100.00
PSO-FNN	0.000585 ± 0.000749	100.00
GA-FNN	5.08e-24 ± 1.06e-23	100.00
ACO-FNN	0.004854 ± 0.007760	100.00
ES-FNN	0.019055 ± 0.170260	100.00
PBIL-FNN	2.49e-05 ± 5.27e-05	100.00

Table 6.6 Experimental results for the iris dataset

Algorithm	MSE (AVE ± STD)	Classification rate (%)
GWO-FNN	0.0229 ± 0.0032	91.333
PSO-FNN	0.228680 ± 0.057235	37.33
GA-FNN	0.089912 ± 0.123638	89.33
ACO-FNN	0.405979 ± 0.053775	32.66
ES-FNN	0.314340 ± 0.052142	46.66
PBIL-FNN	0.116067 ± 0.036355	86.66

6.4.3 Iris Dataset

The Iris dataset is indeed one of the most popular datasets in the literature. It consists of 4 attributes, 150 training/test samples, and 3 classes. Therefore, the FNN structure for solving this dataset is of 4-9-3 and the problem has 75 variables. The results of training algorithms are presented in Table 6.6.

This Table shows that the GWO-FNN outperforms other algorithms in terms on not only the minimum MSE but also the maximum classification accuracy. The results of GWO-FNN follow by those of GA-FNN and PBIL-FNN. Since the difficulty of this dataset and FNN structure is high for this dataset, these results are strong evidences for the efficiencies of GWO in training FNNs. The results testify that this algorithm has superior local optima avoidance and accuracy simultaneously.

6.4.4 Breast Cancer Dataset

This dataset is another challenging well-known dataset in the literature. There are 9 attributes, 599 training samples, 100 test samples, and 2 classes in this dataset. As may be seen in Table 3, an FNN of 4-9-1 is chosen to be trained and solve this dataset. Therefore, every search agent of the trainers in this work have 209 variables to be

Table 6.7 Experimental results for the breast cancer dataset

Algorithm	MSE (AVE ± STD)	Classification rate (%)
GWO-FNN	0.0012 ± 7.4498e-05	99.00
PSO-FNN	0.034881 ± 0.002472	11.00
GA-FNN	0.003026 ± 0.001500	98.00
ACO-FNN	0.013510 ± 0.002137	40.00
ES-FNN	0.040320 ± 0.002470	6.00
PBIL-FNN	0.032009 ± 0.003065	7.00

optimised. This dataset is solved by the trainers 10 times, and the statistical results are provided in Table 6.7.

The results of this table are consistent with those of Table 6.6, in which the GWO-FNN algorithm again shows the best results. The average and standard deviation provided by GWO-FNN proves that this algorithm has a high ability to avoid local minima and approximates the best optimal values for weights and biases. The second best results belong to GA-FNN and PBIL-FNN algorithms. In addition to MSEs, the best classification accuracy of the proposed GWO-FNN algorithm is higher than others. The breast cancer dataset has the highest difficulty compared to the previously discussed datasets in terms of the number of weights, biases, and training samples. Therefore, these results strongly evidence the suitability of the proposed GWO-FNN algorithm in training FNNs. For one, this algorithm shows high local optima avoidance. For another, the local search around to global optimum and exploitation are high.

It should be noted here that the classification rates of some of the algorithms are very small for this dataset because the end criterion for all algorithms is the maximum number of iteration. Moreover, the number of search agents is fixed throughout the experiments. Of course, increasing the number of iterations and population size would improve the absolute classification rate but it was the comparative performance between algorithms that was of interest. It was the ability of algorithms in terms of avoiding local minima in the classification that was the main objective of this work, so the focus was not on finding the best maximum number of iterations and population size.

6.4.5 Heart Dataset

The last classification dataset solved by the algorithms is the heart dataset, which has 22 features, 80 training samples, 187 test samples, and 2 classes. FNNs with the structure of 22-45-1 are trained by the algorithms. The results are reported in Table 6.8.

The results of this table reveal that GWO-FNN and GA-FNN have the best performances in this dataset in terms of improved MSE. The average and standard deviation

Table 6.8 Experimental
results for the heart dataset

Algorithm	MSE (AVE ± STD)	Classification rate (%)
GWO-FNN	0.122600 ± 0.007700	75.00
PSO-FNN	0.188568 ± 0.008939	68.75
GA-FNN	0.093047 ± 0.022460	58.75
ACO-FNN	0.228430 ± 0.004979	0.00
ES-FNN	0.192473 ± 0.015174	71.25
PBIL-FNN	0.154096 ± 0.018204	45.00

of MSEs show that the performances of these two algorithms are very close. However, the classification accuracy of the GWO-FNN algorithm is much higher. Once more, these results evidence the merits of the proposed GWO-based algorithm in training FNNs.

6.4.6 Discussion and Analysis of the Results

Statistically speaking, the GWO-FNN algorithm provides superior local optima avoidance in six of the datasets (75%) and the best classification accuracy in all of the datasets (100%). The reason for improved MSE is the high local optima avoidance of this algorithm. According to the mathematical formulation of the GWO algorithm, half of the iterations are devoted to exploration of the search space (when $|A| > 1$). This promotes exploration of the search space that leads to finding diverse FNN structures during optimisation. In addition, the C parameter always randomly obliges the search agents to take random steps towards/outwards the prey. This mechanism is very helpful for local optima stagnation even when the GWO algorithm is in the exploitation phase. The results of this work show that although evolutionary algorithms have high exploration, the problem of training an FNN needs high local optima avoidance during the whole optimisation process. This is because the search space is changed for every dataset in training FNNs. The results prove that the GWO is very effective in this regard.

Another finding in the results is the poor performance of PSO-FNN and ACO-FNN. These two algorithms belong to the class of swarm-based algorithms. In contrary to evolutionary algorithms, there is no mechanism for abrupt movements in the search space and this is likely to be the reason for the poor performance of PSO-FNN and ACO-FNN. Although GWO is also a swarm-based algorithm, its mechanisms described in the preceding paragraph are why it is advantageous in training FNNs.

It is also worth discussing the poor performance of the ES algorithm in this subsection. Generally speaking, the ES algorithm has been designed based on various mutation mechanisms. Mutation in evolutionary algorithms maintains the diversity of population and promotes exploitation, which is one of the main reasons for the poor performance of ES. In addition, selection of individuals in this algorithm is done by a deterministic approach. Consequently, the randomness in selecting an individual is less and therefore local optima avoidance is less as well. This is another reason why EA failed to provide good results in the datasets.

The reason for the high classification rate provided by the GWO-FNN algorithm is that this algorithm is equipped with adaptive parameters to smoothly balance exploration and exploitation. Half of the iteration is devoted to exploration and the rest to exploitation. In addition, the GWO algorithm always saves the three best obtained solutions at any stage of optimisation. Consequently, there are always guide search agents for exploitation of the most promising regions of the search space. In other words, GWO-FNN benefits from intrinsic exploitation guides, which also assist this algorithm to provide remarkable results.

According to this comprehensive study, the GWO algorithm is highly recommended to be used in hybrid intelligent optimisation schemes such as training FNNs. Firstly, this recommendation is made because of its high exploratory behaviour, which results in high local optima avoidance when training FNNs. The high exploitative behaviour is another reason why a GWO-based trainer is able to converge rapidly towards the global optimum for different datasets. However, it should be noted here that GWO is highly recommended only when the dataset and the number of features are very large. Obviously, small datasets with very few features can be solved by gradient-based training algorithms much faster and without extra computational cost. In contrast, the GWO algorithm is useful for large datasets due to the extreme number of local optima that makes the conventional training algorithm almost ineffective.

6.5 Conclusion

In this chapter, several evolutionary trainers were introduced to train FNNs using several conventional and recent evolutionary and swarm-based algorithms. The problem of training an FNN was first formulated for the algorithms. These algorithm were then employed to define the optimal values for weights and biases. The proposed GWO-based trainers were applied to five standard classification datasets (XOR, balloon, Iris, breast cancer, and heart) and the results of the GWO-FNN algorithm were compared to five other stochastic optimisation trainers: PSO, GA, ACO, ES, and PBIL. The results showed that the GWO-FNN has very a high level of local optima avoidance, which enhances the probability of finding accurate approximations of the optimal weights and biases for FNNs. Moreover, the accuracy of the obtained optimal values for weights and biases is very high, which is due to the high exploitation of the GWO-FNN trainer. This chapter also identified and discussed the reasons for strong and poor performances of other algorithms. It was observed that the swarm-based algorithms suffer from low exploration as opposed to evolutionary algorithms.

References

1. Magoulas, G. D., Vrahatis, M. N., & Androulakis, G. S. (1997). On the alleviation of the problem of local minima in back-propagation. *Nonlinear Analysis, 30*(7), 4545–4550.
2. Bebis, G., & Georgiopoulos, M. (1994). Feed-forward neural networks. *IEEE Potentials, 13*(4), 27–31.
3. Moallem, P., & Razmjooy, N. (2012). A multi layer perceptron neural network trained by invasive weed optimization for potato color image segmentation. *Trends in Applied Sciences Research, 7*(6), 445.
4. Mirjalili, S., Mirjalili, S. M., & Lewis, A. (2014). Grey wolf optimizer. *Advances in Engineering Software, 69*, 46–61.
5. Nan, W., Wenxiao, S., Shaoshuai, F., & Shuxiang, L. (2011). PSO-FNN-based vertical handoff decision algorithm in heterogeneous wireless networks. *Procedia Environmental Sciences, 11*, 55–62.
6. Belew, R. K., McInerney, J., & Schraudolph, N. N. (1990). Evolving networks: Using the genetic algorithm with connectionist learning. In C. G. Langton, C. Taylor, J. D. Farmer, & S. Rasmussen (Eds.), *Artificial Life II, SFI Studies in the Sciences of Complexity: Proceedings* (pp. 511–547). Redwood City, CA: Addison-Wesley.
7. Mei, H., & Wang, Y. (2009). Ant Colony Optimization for Neural Network. In *Key engineering materials* (Vol. 392, pp. 677–681). Trans Tech Publications.
8. Beyer, H. G., & Schwefel, H. P. (2002). Evolution strategies a comprehensive introduction. *Natural Computing, 1*(1), 3–52.
9. Baluja, S. (1994). Population-based incremental learning. a method for integrating genetic search based function optimization and competitive learning (No. CMU-CS-94-163). Carnegie-Mellon Univ Pittsburgh Pa Dept Of Computer Science.
10. Merz, C. J., & Murphy, P. M. (1998). UCI Repository of machine learning databases.
11. Mirjalili, S. (2015). How effective is the Grey Wolf optimizer in training multi-layer perceptrons. *Applied Intelligence, 43*(1), 150–161.

Chapter 7
Evolutionary Multi-layer Perceptron

7.1 Introduction

One of the more significant inventions in the field of soft computing is Neural Networks (NN), inspired by biological neurons in the human brain. The rudimentary concepts of NN were first mathematically modelled by McCulloch and Pitts [1]. The simplicity, low computational cost, and high performance have made this computational tool remarkably popular over the last decade. Among different types of NNs, the Feedforward Neural Network (FNN) [2] is the simplest and most widely-used.

FNNs receive information as inputs on one side and provide outputs from the other side using one-directional connections between the neurons in different layers. There are two types of FNN: Single-Layer Perceptron (SLP) [3] and Multi-Layer Perceptron (MLP) [4]. In the SLP there is only a single perceptron that makes it suitable for solving linear problems. However, an MLP has more than one perceptron, established in different layers. This makes it capable of solving non-linear problems.

Generally speaking, the applications of MLPs are categorized as pattern classification [5], data prediction [6], and function approximation [7]. Pattern classification implies classifying data into pre-defined discrete classes [8], whereas prediction refers to the forecasting of future trends according to current and previous data [6]. Finally, function approximation involves the process of modelling relationships between input variables. It has been proven that MLPs with one hidden layer are able to approximate any continuous or discontinuous functions [9, 10]. Regardless of the applications, the distinguishing capability of MLPs is learning [1]. MLPs are equipped with a learning concept that gives them the ability to learn from experience, similar to a human brain. This component is an essential part of all NNs. It may be divided into two types: supervised and unsupervised learning. For MLPs, most applications use the standard [11] or improved Back-Propagation (BP) [12] algorithms as their learning methods, which belong to the supervised learning fam-

Part of this chapter has been reprinted from Seyedali Mirjalili, Seyed Mohamed Mirjalili and Andrew Lewis article: Let a biogeography-based optimizer train your multilayer perceptron, Information Science, Volume 269, pp. 188–209, 2014

S. Mirjalili, *Evolutionary Algorithms and Neural Networks*, Studies
in Computational Intelligence 780, https://doi.org/10.1007/978-3-319-93025-1_7

ily. Back Propagation (BP) is a gradient-based algorithm that has some drawbacks such as slow convergence [13] and a tendency to get trapped in local minima [14], making it unreliable for practical applications.

The ultimate goal of the learning process is to find the best combination of connection weights and biases in the NN to achieve the minimum error for training and test samples. However, often the error of MLP stays constantly large for some extended period of time during the learning process, as the learning algorithm leads MLPs to local minima rather than the global minimum. This problem is quite common in gradient-based learning approaches such as BP. The convergence of BP is also highly dependent on the initial values of learning rate and momentum. Unsuitable values for these variables may even result in divergence. There are many studies focused on resolving these problems of BP (e.g. [15]), but there is no reported significant improvement, and each method has its own side effects. The literature shows that heuristic optimisation methods are promising alternatives for gradient-based learning algorithms [16] since the stochastic nature of these algorithms allows them to avoid local minima better than gradient-based techniques and optimise challenging problems. Moreover, convergence rates of heuristic methods to the global minimum can be faster than BP, as investigated in [17].

7.2 Multi-layer Perceptrons

In MLP formulation, it is normally assumed that there are n number of input neurons, the number of hidden nodes is h, and the number of output nodes is m. There are one-way connections between the nodes, since the MLP belongs to the FNN family. The output of the MLP is calculated as follows:

The weighted sums of inputs are first calculated by Eq. 7.1.

$$s_j = \sum_{i=1}^{n} \left(W_{ij} X_i \right) - \theta_j, \qquad j = 1, 2, \ldots, h \tag{7.1}$$

where n is the number of the input nodes, W_{ij} shows the connection weight from the ith node in the input layer to the jth node in the hidden layer, θ_j is the bias (threshold) of the jth hidden node, and X_i indicates the ith input.

The output of each hidden node is calculated as follows:

$$S_j = sigmoid(s)j) = \frac{1}{1 + e^{-s_j}}, \qquad j = 1, 2, \ldots, m \tag{7.2}$$

After calculating the outputs of hidden nodes, the final outputs are defined as follows:

$$o_k = \sum_{j=1}^{h} \left(w_{jk} S_j \right) - \theta'_k, \qquad k = 1, 2, \ldots, m \tag{7.3}$$

$$O_k = sigmoid(o_k) = \frac{1}{1 + e^{-o_k}}, \quad k = 1, 2, \ldots, m \quad (7.4)$$

where w_{jk} is the connection weight from the jth hidden node to the kth output node, and $'_k$ is the bias (threshold) of the kth output node.

The most important parts of MLPs are the connection weights and biases. As may be seen in the above equations, the weights and biases define the final values of output. Training an MLP involves finding optimum values for weights and biases in order to achieve desirable outputs from certain given inputs.

7.3 Evolutionary Multi-layer Percenptron

Generally, there are three methods of using a heuristic algorithm for training MLPs. Firstly, heuristic algorithms are utilized for finding a combination of weights and biases that provide the minimum error for an MLP. Secondly, heuristic algorithms are employed to find a proper architecture for an MLP in a particular problem. The last method is to use a heuristic algorithm to tune the parameters of a gradient-based learning algorithm, such as the learning rate and momentum.

In the first method, the architecture does not change during the learning process. The training algorithm is required to find proper values for all connection weights and biases in order to minimise the overall error of the MLP. In the second approach, the structure of the MLPs varies. In this case, a training algorithm determines the best structure for solving a certain problem. Changing the structure can be accomplished by manipulating the connections between neurons, the number of hidden layers, and the number of hidden nodes in each layer. For example, Yu et al. employed PSO to define the structure of MLP to solve two real problems [18].

Leung et al. used the last method to tune the parameters of an FNN utilizing EAs [19]. There are also some studies that utilized a combination of methods simultaneously. For instance, Mizuta et al. [20] and Leung et al. [19] employed a GA and improved GA to define the structure of an FNN.

In the next section, a variety of EAs are employed to train several MLPs for function approximation.

7.4 Results

In this section the EA algorithms are benchmarked on six function approximation datasets. The function approximation datasets are a one-dimensional sigmoid, one-dimensional cosine with one peak, one-dimensional sine with four peaks, two-dimensional sphere, two-dimensional Griewank, and five-dimensional Rosenbrock functions. The algorithms employed are BBO, PSO, GA, ACO, ES, and PBIL over these benchmark datasets. The comparison of the best WA with BP and ELM are also given.

Table 7.1 Function-approximation datasets

Function-approximation datasets	Training samples	Test samples
Sigmoid: $y = \frac{1}{1+e^{-x}}$	61: x in [-3:0.1:3]	121: x in [-3:0.05:3]
Cosine: $y = cos\left(\frac{x\pi}{2}\right)^7$	31: x in [1.25:0.05:2.75]	38: x in [1.25:0.04:2.75]
Sine: $y = sin(2x)$	126: x in [-2:0.1:2]	252: x in [-2:0.05:2]
Sphere: $z = \sum_{i=1}^{2}, x = x_1, y = x_2$	21*21: x, y in [-2:0.2:2]	41*41: x, y in [-2:0.1:2]
Griewank: $z = \frac{1}{4000}\sum_{i=1}^{2} x_i^2 cos\left(\frac{x_i}{\sqrt{i}}+1, x = x_1, y = x_2\right)$	21*21: x, y in [-1:0.1:1]	41*41: x, y in [-1:0.05:1]
Rosenbrock: $z = \sum_{i=1}^{5}\left[x_i^2 - 10cos(2\pi x_i) + 10\right]$	1024: $x_1 - x_5$ in [-5:3:5]	10247776: $x_1 - x_5$ in [-5:2:5]

It is assumed that the weights and biases are randomly initialized in the range
$[-10, 10]$. The population size is 200 for all function approximation datasets.
Table 7.1 show hows the datasets are divided in terms of training and test sets. It
can be seen that the training and test samples for the function approximation datasets
are chosen using different step sizes from the domains of the functions. In most of
the function approximation datasets the test samples are twice the training samples.

The other assumptions and initial values are presented in Table 7.2. Fine-tuning
of the algorithms is beyond the scope of this paper. Each algorithm was run 10 times
and the average (AVE) and standard deviation (STD) are reported in the table of
results. These two measures indicate the ability of algorithms to avoid local minima.
The lower the value of AVE ± STD, the greater the capability of the algorithm to
avoid local minima. AVE indicates the average of MSE over 10 runs, so a lower
value for this metric is evidence of an algorithm more successfully avoiding local
optima and finding solutions near the global optimum. However, AVE is not a good
metric alone because two algorithms can have equal averages, but have different
performance in terms of finding the global optimum in each run. Therefore, STD
(standard deviation) can help to determine the dispersion of results. The lower the
STD, the lower the dispersion of results. So, AVE±STD may be a good combination
to indicate the performance of an algorithm in terms of avoiding local minima.

According to Derrac et al. [21], statistical tests should be conducted to properly
evaluate the performance of heuristic algorithms. It is not enough to compare algorithms based on the mean and standard deviation values; a statistical test is necessary
to prove that a proposed new algorithm presents a significant improvement over other
existing methods for a particular problem [22].

In order to judge whether the results of the best EA are better in a statistically
significant way, a nonparametric statistical test, Wilcoxon's rank-sum test [23], was
carried out at 5% significance level. The calculated p-values in the Wilcoxon's ranksum are given in the results as well. In the tables, N/A indicates Not Applicable which
means that the corresponding algorithm cannot be compared with itself in the ranksum test. Conventionally, p-values less than 0.05 are considered as strong evidence
against the null hypothesis. Note that p-values greater than 0.05 are underlined in
the tables.

Table 7.2 Initial values for the controlling parameters of EAs

Algorithm	Parameter	Value
BBO	Habitat modification probability	1
	Immigration probability bounds per gene	[0, 1]
	Step size for numerical integration of probabilities	1
	Max immigration (I) and Max emigration (E)	1
	Mutation probability	0.005
	Population size	50 for XOR and Balloon, 200 for the rest
	Maximum number of generations	250
PSO	Topology	connected
	Cognitive constant (c_1)	1
	Social constant (c_2)	1
	Inertia constant (w)	0.3
	Population size	50 for XOR and Balloon, 200 for the rest
	Maximum number of iterations	250
GA	Type	Real coded
	Selection	Roulette wheel
	Crossover	Single point (probability=1)
	Mutation	Uniform (probability=0.01)
	Population size	50 for XOR and Balloon, 200 for the rest
	Maximum number of generations	250
ACO	Initial pheromone (τ_0)	1e-06
	Pheromone update constant (Q)	20
	Pheromone constant (q_0)	1
	Global pheromone decay rate (p_g)	0.9
	Local pheromone decay rate (p_t)	0.5
	Pheromone sensitivity (α)	1
	Visibility sensitivity (β)	5
	Population size	50 for XOR and Balloon, 200 for the rest
	Maximum number of iterations	250
ES	λ	10
	σ	1
	Population size	50 for XOR and Balloon, 200 for the rest
	Maximum number of generations	250
PBIL	Learning rate	0.05
	Good population member	1
	Bad population member	0
	Elitism parameter	1
	Mutational probability	0.1
	Population size	50 for XOR and Balloon, 200 for the rest
	Maximum number of generations	250

The other comparative measures shown in the results are: classification rates and test errors. The best trained MLP among 10 runs is chosen to approximate the test set. To provide a fair comparison, all algorithms were terminated when a maximum number of iterations (250) was reached. Finally, the convergence behaviour is also investigated in the results to provide a comprehensive comparison.

It should be noted that min-max normalization was used for those datasets containing data with different ranges. The normalization method was formulated as follows: Suppose that we are going to map x in the interval of $[a, b]$ to $[c, d]$. The normalization process is done by the following equation:

$$x' = \frac{(x - a) \times (d - c)}{(b - a)} + c \qquad (7.5)$$

Regarding the structure of the MLPs, 15 hidden nodes are used for function approximation datasets. In the following sections the simulation results of benchmark datasets are presented and discussed [24].

7.4.1 Sigmoid Function

The sigmoid dataset is in the interval $[-3, 3]$ with increments of 0.1, so the number of training data is 61. The number of test samples is 121, lying in the same range. The results of training algorithms for this dataset are presented in Table 7.3, Figs. 7.1 and 7.2. The results for AVE, STD, and p-values indicate that BBO is much better at avoiding local minima than the other algorithms. The test errors in Table 7.3 and approximated curves in Fig. 7.1 show that the BBO algorithm has the best approximate accuracy as well. Figure 7.2 shows that BBO has the fastest convergence rate.

7.4.2 Cosine Function

This dataset has 31 training samples and 38 test samples. The results are shown in Table 7.4. It can be seen that the GA has the best results for AVE, STD, and test errors. However, the results of BBO are very close to this: the statistical tests show that the differences between the GA and BBO are not statistically significant. Figures 7.3 and 7.4 illustrate the approximated curve and convergence, respectively.

Table 7.3 Experimental results for sigmoid dataset (one dimensional)

Algorithm	MSE (AVE ± STD)	P-values	Test error
BBO	1.33e-05 ± 3.57e-21	N/A	0.14381
PSO	0.022989 ± 0.009429	6.39E-05	3.3563
GA	0.001093 ± 0.000916	6.39E-05	0.44969
ACO	0.023532 ± 0.010042	6.39E-05	3.9974
ES	0.075575 ± 0.016410	6.39E-05	8.8015
PBIL	0.004046 ± 2.74e-17	6.34E-05	2.9446

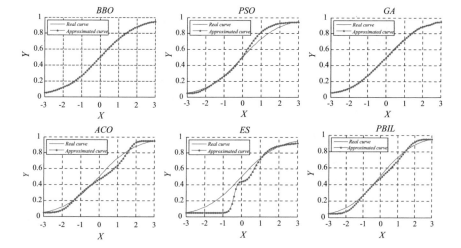

Fig. 7.1 Approximated curves for sigmoid function

Fig. 7.2 Convergence curves of algorithms for sigmoid function

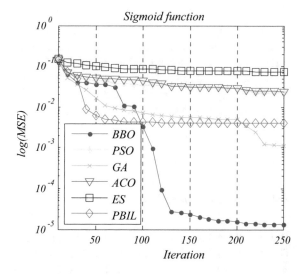

Table 7.4 Experimental results for one-peak cosine dataset (one dimensional)

Algorithm	MSE (AVE ± STD)	P-values	Test error
BBO	0.013674 ± 1.83e-18	0.4429	1.4904
PSO	0.058986 ± 0.021041	0.0001	2.009
GA	0.010920 ± 0.006316	N/A	0.7105
ACO	0.050872 ± 0.010809	0.0001	2.4498
ES	0.086640 ± 0.022208	0.0001	3.1461
PBIL	0.094342 ± 0.018468	0.0001	3.727

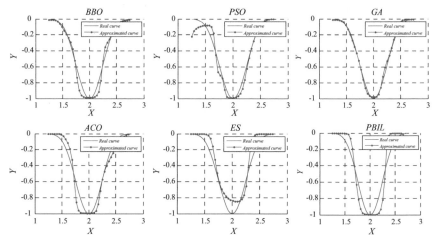

Fig. 7.3 Approximated curves for cosine function

7.4.3   Sine Function

This dataset is created by a four-peak sine function in the interval $[-2, 2]$ with a step size of 0.1 for training samples and 0.05 for test samples. For training the algorithms use 126 samples and are tested by 252 samples. The experimental results are provided in Table 7.5, Figs. 7.5 and 7.6. The results show that the BBO algorithm is significantly better than the other algorithms in terms of avoiding local minima. In addition, test errors and Fig. 7.5 verify the superior accuracy of this algorithm. Finally, the curves of Fig. 7.6 confirm that BBO provides the fastest convergence speed.

7.4.4   Sphere Function

The sphere function dataset consists of 441 training samples and 1681 test samples. The results of this dataset are reported in Table 7.6. The averages and standard deviations of the MSEs show that BBO is the most reliable algorithm in terms of

Table 7.5 Experimental results for four-peak sine dataset (one dimensional)

Algorithm	MSE (AVE ± STD)	P-values	Test error
BBO	0.102710 ± 0.000000	N/A	64.261
PSO	0.526530 ± 0.072876	6.39E-05	124.89
GA	0.421070 ± 0.061206	6.39E-05	111.25
ACO	0.529830 ± 0.053305	6.39E-05	117.71
ES	0.706980 ± 0.077409	6.39E-05	142.31
PBIL	0.483340 ± 0.007935	6.39E-05	149.6

avoiding local minima. However, the results for the test error criterion and the best approximation surfaces in Fig. 7.7 indicate that the GA provides better accuracy. The convergence behaviours of the training algorithm are shown in Fig. 7.8. The convergence rates are entirely consistent with the previous datasets: BBO has the fastest rate, followed by the GA.

Fig. 7.4 Convergence curves of algorithms for cosine function

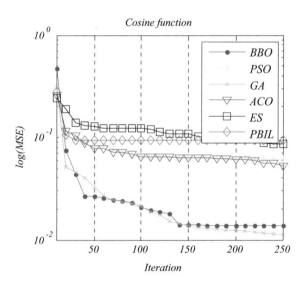

Table 7.6 Experimental results for sphere dataset (two dimensional)

Algorithm	MSE (AVE ± STD)	P-values	Test error
BBO	1.7740 ± 2.34e-16	N/A	770.4425
PSO	7.1094 ± 0.8528	6.38E-05	1.27E+03
GA	2.9276 ± 0.7289	6.38E-05	452.3744
ACO	6.5626 ± 0.9543	6.38E-05	1.22E+03
ES	10.530 ± 0.8427	6.38E-05	1.50E+03
PBIL	10.459 ± 0.8529	6.38E-05	1.81E+03

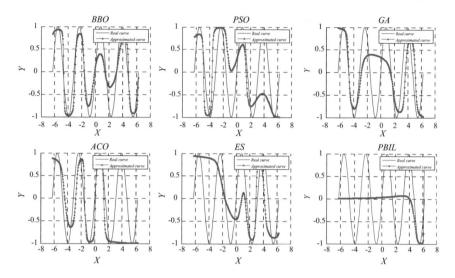

Fig. 7.5 Approximated curves for sin function

Fig. 7.6 Convergence
curves of algorithms for sin
function

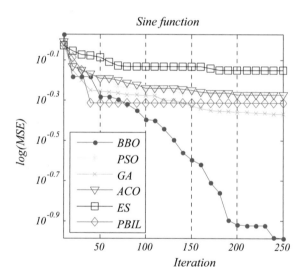

7.4.5 Griewank Function

The Griewank dataset includes 441 training instances and 1681 test samples. The
results for this dataset are shown in Table 7.7, Figs. 7.9 and 7.10. The results are quite
similar to the previous datasets in that the BBO algorithm significantly outperforms
the other algorithms in terms of avoiding local minima. According to test errors,

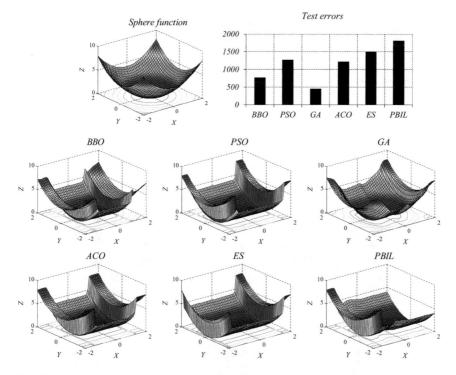

Fig. 7.7   Approximated curves for the sphere function

Fig. 7.8   Convergence
curves of algorithms for the
sphere function

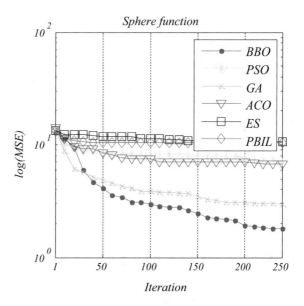

Table 7.7 Experimental results for Griewank dataset (two dimensional)

Algorithm	MSE (AVE ± STD)	P-values	Test error
BBO	2.02560 ± 0.0000	N/A	5.02E+03
PSO	16.9043 ± 1.1482	6.38E-05	7.63E+03
GA	16.4218 ± 1.3306	6.38E-05	1.06E+04
ACO	14.9586 ± 1.2976	6.38E-05	6.18E+03
ES	19.1069 ± 1.4984	6.38E-05	1.12E+04
PBIL	13.5587 ± 0.7532	6.29E-05	1.24E+04

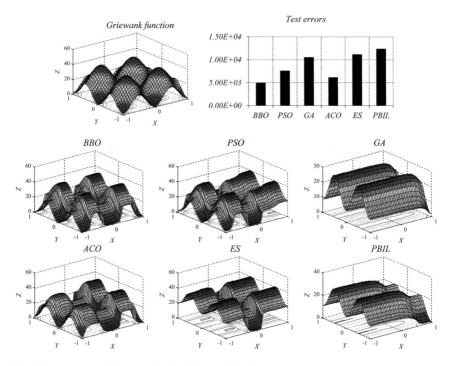

Fig. 7.9 Approximated curves for the Griewank function

Figs. 7.9 and 7.10, the BBO algorithm has the best approximation accuracy and convergence speed as well.

7.4.6 Rosenbrock Function

This dataset includes 1024 training samples and 7776 test samples. The results for this dataset are provided in Table. 7.8. According to the values of AVE, STD, and p-

Fig. 7.10 Convergence
curves of algorithms for the
Griewank function

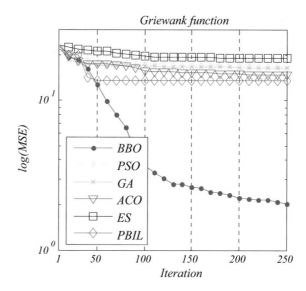

Table 7.8 Experimental results for Rosenbrock dataset (five-dimensional)

Algorithm	MSE (AVE ± STD)	P-values	Test error
BBO	104.4251 ± 1.59e-14	N/A	2.35E+06
PSO	215.1669 ± 3.5521	0.007937	6.50E+09
GA	264.6713 ± 16.3603	0.007937	5.24E+06
ACO	200.9749 ± 7.0641	0.007937	3.54E+07
ES	252.8454 ± 8.1445	0.007937	7.37E+09
PBIL	208.6148 ± 4.6813	0.007937	2.35E+08

values, the BBO algorithm significantly outperforms others in terms of avoiding local minima for this dataset. The test errors show that BBO has the highest accuracy for approximating the Rosenbrock function. Figure 7.11 shows that the BBO algorithm has the fastest convergence speed.

Statistically speaking, the BBO algorithm provides significant ability to avoid local minima and a high convergence rate on four out of six function approximation datasets. Moreover, the approximated functions using BBO are more accurate than the other training algorithms on four out of six function approximation datasets. The algorithm shows improved ability to avoid local minima.

Note that the test errors of all algorithms are not very small for all function approximation datasets because the end criterion for all algorithms is the maximum number of iterations and the number of search agents is fixed throughout the experiments. Of course, increasing the number of iterations and population size would improve the absolute classification rate or test error but it was the comparative performance between algorithms that was of interest. Determining the ability of algorithms in

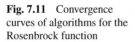
Fig. 7.11 Convergence curves of algorithms for the Rosenbrock function

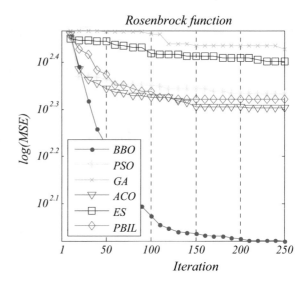

terms of avoiding local minima in the classification or approximation search spaces was the main objective of this work, so we did not focus on finding the best maximum number of iterations and population size.

One of the things that can be inferred from the results is the superior performance of BBO and the GA in all search spaces. This is possibly due to the nature of these algorithms which have migration and crossover strategies to avoid local minima. These operators cause abrupt changes in the candidate solutions that results in enhancing the exploration ability of BBO and the GA significantly. Since the problem of training MLPs has a very difficult search space and is changed for every dataset, these algorithms performed very well in this work because of their very good exploration. The poor results of ACO and PSO are also interesting, originating from the nature of these problems. ACO and PSO do not have operators that promote sudden changes in the candidate solutions, so they are trapped in local minima more often than BBO and the GA. Moreover, ACO uses a pheromone matrix that enhances reinforcement learning and exploitation, an advantage for combinatorial problems but again leading to a tendency toward local minima entrapment. The PSO algorithms are also highly dependent on the distribution of the initial swarm, and their prime motivators are based on attraction between members of the swarm. If many of the particles become trapped in a wide local optimum, there is little in the algorithm to prevent other particles from also being attracted and becoming trapped in that same local optimum.

Another fact worth mentioning in the results is the poor performance of ES, especially on the function approximation datasets. The ES algorithm mainly relies on sophisticated mutation operators and the selection procedure of the individuals is deterministic compared to the GA. Consequently, the mechanism of selection and mutation mostly emphasizes exploitation rather than exploration which results in very poor outcomes on the datasets. The PBIL algorithm also performed very badly on most of the datasets. PBIL evolves a vector consisting of the entire population rather than each individual itself. This vector is updated based on the fittest individual and mutated randomly. This mechanism appears to provide less exploration than the GA, where the production of new individuals by crossover operators, introducing bulk changes in their structure, enhances exploration. This may be the main reason of PBILs poor performance.

The reason for superior performance of the BBO algorithm compared to GA in the majority of the datasets is due to the various emigration and immigration rates for habitats. In contrast to the GA that has a general reproduction rate for all individuals, the BBO algorithm assigns two rates (emigration and immigration) to each habitat, which results in it providing different evolutionary behaviours and eventually greater exploration. As the results of Table 7.3–7.8 show, the significant superiority of the BBO algorithm is increased as the complexity of the data sets increases. This is again due to diverse evolutionary mechanisms (greater exploration) of the BBO algorithm that assists this algorithm to outperform the GA algorithm with a unified mechanism for reproduction. The superior performance of the BBO algorithm compared to PSO and ACO is due to the migration mechanism of this algorithm. The PSO and ACO algorithms are not equipped with an evolutionary operator, so they do not provide sudden changes in the candidate solutions, as discussed above. The BBO algorithm, however, promotes diverse abrupt change mechanisms for candidate solutions, making this algorithm able to outperform PSO and ACO in almost all of the datasets.

The superiority of BBO compared to other evolutionary algorithms employed in this paper (EA and PBIL) can be reasoned from both the exploration and exploitation mechanisms. The selection process of the EA algorithm is deterministic, and the main characteristic is sophisticated mutation operators. These two facts promote exploitation. The BBO algorithm, however, promotes exploration and exploitation simultaneously. On one hand, the various migration rates assist BBO with greater exploration. On the other hand, the different mutation rates for each habitat allow BBO to use different exploitation behaviours for every habitat. The results show that BBO can outperform EA not only in terms of local minima avoidance but also convergence rate. Comparing the results of BBO and PBIL, we can come to the same conclusion. As discussed above, the PBIL evolves a vector consisting of the entire population rather than each individual itself. This vector is updated based on the fittest individual and mutated randomly. This mechanism appears to provide less exploration than BBO, where the production of new habitats by emigration and immigration operators enhances exploration. This is the main reason of BBOs significantly better performance.

To sum up, it can be stated that exploration is very important in the problem of training MLPs. There is a need to have more random and abrupt search steps (emphasizing exploration) to avoid local minima for solving complex data sets using MLPs. This study shows that the operators of BBO (migration) are highly suitable for addressing this issue.

7.4.7 Comparison with BP

To further investigate the efficiency of the BBO algorithm in training MLPs, a comparative study between this algorithm and the BP algorithm on all the datasets is provided in the following section.

The BP algorithm is a gradient-based algorithm that uses gradient information for minimising the error function. We chose this algorithm for comparison because it is totally based on mathematical concepts, quite different from the other, bio-inspired algorithms. This can benchmark the performance of the proposed approach on a contrasting test bed. We solved the datasets using the BP algorithm 10 times and provide the statistical results in Table 7.9, as well as those of BBO. Note that the learning rate, momentum, and maximum number of iterations are set to 0.01, 0.001, and 250 respectively. It can be seen that the results of the BBO algorithm for AVE and STD are better than BP in most of the datasets. This shows the superior performance of BBO in terms of improved avoidance of local minima. The classification rates and test errors also show that BBO is better than BP in most of the datasets. However, the BP algorithm provided very competitive results for simple datasets. This shows that the performance of BP is degraded as the complexity of the search space is increased (due to the problem of entrapment in local minima).

The reasons for the better performance of BBO compared to BP and other heuristic algorithms are as follows:

- Varying values of emigration and immigration rates provide diverse information exchange behaviour between habitats, and consequently improve exploration.
- Over the course of generations, the HSI of all habitats are improved since habitants living in high-HSI habitats tend to migrate to the low-HSI habitats. This guarantees the convergence of BBO.
- Migration operators emphasize exploration and consequently prevent BBO from easily getting trapped in local minima.
- Different mutation rates keep habitats as diverse as possible.

Table 7.9 Comparison results between BBO and BP

Dataset	Algorithm	MSE (AVE ± STD)	P-values	Test error
Sigmoid	BBO	1.33e-05 ± 3.57e-21	6.3864e-005	0.14381
	BP	3.70e-04 ± 1.26e-04		1.3894
Cosine	BBO	0.013674 ± 1.83e-18	1.8267e-004	1.4904
	BP	0.007932 ± 0.003799		2.5663
Sine	BBO	0.102710 ± 0.000000	6.3864e-005	64.261
	BP	0.020270 ± 0.00745		23.3423
Sphere	BBO	1.774000 ± 2.34e-16	1.8267e-004	770.4425
	BP	1.544630 ± 0.396995		1.3618e+03
Griewank	BBO	2.025600 ± 0.0000	6.3864e-005	5.0200e+03
	BP	3.325595 ± 6.1857		9.8313e+003

- Elitism assists BBO to save and retrieve some of the best solutions, so they are never-lost. In the following section the BBO algorithm is compared to a very effective method called ELM.

According to this comparative study and discussion, it appears that BBO is highly suitable for training MLPs. This algorithm successfully reduces the problem of entrapment in local minima in training MLPs, with very fast convergence rates. These improvements are accompanied by high classification rates and low test errors as well.

7.5 Conclusion

In this chapter, an MLP was trained by six evolutionary algorithms. Six benchmark datasets (sigmoid, cosine with one peak, sine with four peaks, sphere, Griewank, and Rosenbrock) were employed to investigate the effectiveness of BBO in training MLPs. The results were statistically compared with PSO, GA, ACO, ES, PBIL, BP, and ELM to verify the performance. The results demonstrate that BBO is significantly better able to avoid local minima compared to PSO, GA, ACO, ES, and PBIL. Moreover, the superior performance of BBO for training MLPs in terms of accuracy of results and convergence rate can clearly be seen from the results.

References

1. McCulloch, W. S., & Pitts, W. (1943). A logical calculus of the ideas immanent in nervous activity. *The Bulletin of Mathematical Biophysics, 5*(4), 115–133.
2. Fahlman, S. E. (1988). An empirical study of learning speed in back-propagation networks.
3. Raudys, Š. (1998). Evolution and generalization of a single neurone: I. Single-layer perceptron as seven statistical classifiers. *Neural Networks, 11*(2), 283–296.
4. Amendolia, S. R., Cossu, G., Ganadu, M. L., Golosio, B., Masala, G. L., & Mura, G. M. (2003). A comparative study of k-nearest neighbour, support vector machine and multi-layer perceptron for thalassemia screening. *Chemometrics and Intelligent Laboratory Systems, 69*(1–2), 13–20.
5. Melin, P., Snchez, D., & Castillo, O. (2012). Genetic optimization of modular neural networks with fuzzy response integration for human recognition. *Information Sciences, 197*, 1–19.
6. Guo, Z. X., Wong, W. K., & Li, M. (2012). Sparsely connected neural network-based time series forecasting. *Information Sciences, 193*, 54–71.
7. Gardner, M. W., & Dorling, S. R. (1998). Artificial neural networks (the multilayer perceptron) a review of applications in the atmospheric sciences. *Atmospheric Environment, 32*(14–15), 2627–2636.
8. Barakat, M., Lefebvre, D., Khalil, M., Druaux, F., & Mustapha, O. (2013). Parameter selection algorithm with self adaptive growing neural network classifier for diagnosis issues. *International Journal of Machine Learning and Cybernetics, 4*(3), 217–233.
9. Csji, B. C. (2001). Approximation with artificial neural networks. *Faculty of Sciences, Etvs Lornd University, Hungary, 24*, 48.
10. Hornik, K., Stinchcombe, M., & White, H. (1989). Multilayer feedforward networks are universal approximators. *Neural Networks, 2*(5), 359–366.

11. Hush, D. R., & Horne, B. G. (1993). Progress in supervised neural networks. *IEEE Signal Processing Magazine, 10*(1), 8–39.
12. Hagan, M. T., & Menhaj, M. B. (1994). Training feedforward networks with the Marquardt algorithm. *IEEE Transactions on Neural Networks, 5*(6), 989–993.
13. Ng, S. C., Cheung, C. C., Leung, S. H., & Luk, A. (2003). Fast convergence for backpropagation network with magnified gradient function. In *2003 Proceedings of the international joint conference on neural networks* (Vol. 3, pp. 1903-1908). IEEE.
14. Lee, Y., Oh, S. H., & Kim, M. W. (1993). An analysis of premature saturation in back propagation learning. *Neural Networks, 6*(5), 719–728.
15. Weir, M. K. (1991). A method for self-determination of adaptive learning rates in back propagation. *Neural Networks, 4*(3), 371–379.
16. Yao, X. (1993). Evolutionary artificial neural networks. *International Journal of Neural Systems, 4*(03), 203–222.
17. Gudise, V. G., & Venayagamoorthy, G. K. (2003). Comparison of particle swarm optimization and backpropagation as training algorithms for neural networks. In *Proceedings of the 2003 swarm intelligence symposium, SIS'03* (pp. 110–117). IEEE.
18. Yu, J., Wang, S., & Xi, L. (2008). Evolving artificial neural networks using an improved PSO and DPSO. *Neurocomputing, 71*(4–6), 1054–1060.
19. Leung, F. H. F., Lam, H. K., Ling, S. H., & Tam, P. K. S. (2003). Tuning of the structure and parameters of a neural network using an improved genetic algorithm. *IEEE Transactions on Neural networks, 14*(1), 79–88.
20. Mizuta, S., Sato, T., Lao, D., Ikeda, M., & Shimizu, T. (2001). Structure design of neural networks using genetic algorithms. *Complex Systems, 13*(2), 161–176.
21. Derrac, J., Garca, S., Molina, D., & Herrera, F. (2011). A practical tutorial on the use of nonparametric statistical tests as a methodology for comparing evolutionary and swarm intelligence algorithms. *Swarm and Evolutionary Computation, 1*(1), 3–18.
22. Mirjalili, S., & Lewis, A. (2013). S-shaped versus V-shaped transfer functions for binary particle swarm optimization. *Swarm and Evolutionary Computation, 9*, 1–14.
23. Wilcoxon, F. (1945). Individual comparisons by ranking methods. *Biometrics Bulletin, 1*(6), 80–83.
24. Mirjalili, S., Mirjalili, S. M., & Lewis, A. (2014). Let a biogeography-based optimizer train your multi-layer perceptron. *Information Sciences, 269*, 188–209.

Chapter 8
Evolutionary Radial Basis Function Networks

8.1 Introduction

Radial Basis Function (RBF) networks are one of the most popular and applied type of neural networks. RBF networks are universal approximators and considered as special form of multilayer feedforward neural networks that contain only one hidden layer with Gaussian based activation functions. RBF networks were first introduced by Lowe and Broomhead in [1] with a strong foundation in the conventional approximation theory [2].

The advantages of RBF networks compared to other neural networks include the high generalization capability, its simple and compact structure (i.e.; only three layers), easier parameters adjustment, very good noise tolerance and the high learning speed [3]. Due to these advantages, RBF networks have been a common alternative to MLP networks [2]. Moreover, RBF networks have been successfully applied to many applications like: systems identification [4], process faults classification [5], nonlinear control [6] and time series forecasting [7].

Like other neural networks, RBF networks have two major components: the structure and the training method. The training method has a significance influence on the performance of the network. In literature, researchers proposed and investigated a wide variety of learning schemes for RFB network.

Castao et al. [8] classified the training methods of RBF networks into two categories: quick learning and full learning. The quick learning methods are more popular where the learning process can be performed in two independent stages: In the first stage, the structure of the network (i.e.; the centers and widths of the network) is identified usually by an unsupervised learning algorithm like K-Means algorithm, while in the second stage, the connection weights between the hidden and output layers are tuned using Least Mean Squares (LMS), gradient based methods and variations of the backpropagation algorithm [9]. The drawback of using an unsupervised technique to locate the centers is that it depends only on the input features and it doesn't consider the distribution of the label class [10]. Moreover, using the common K-Means algorithm doesn't not necessarily guarantee that the centers are best

© Springer International Publishing AG, part of Springer Nature 2019
S. Mirjalili, *Evolutionary Algorithms and Neural Networks*, Studies
in Computational Intelligence 780, https://doi.org/10.1007/978-3-319-93025-1_8

located [11]. On the other side, the main issue with the gradient methods is that its highly probable that the search process will be trapped in a local minima. Moreover, Vakil-Baghmisheh and Pave reported in [12] that applying customized version of the backpropagation algorithm to RBF networks could suffer from some drawbacks like the slow convergence and over-training which consequently affects the generalization ability of the model. Alternatively, the full learning methods optimise the RBF parameters simultaneously as a supervised task.

8.2 Radial Based Function Neural Networks

RBF neural network is a special type of fully connected feedforward networks that consists of only three layers: input, hidden and output layers. The number of neurons in the input layer depends on the number of dimensions of the input vector, whereas output layer neurons depend on the number of class labels in the data. The number of neurons in the hidden layer determines the topology of the network which also determines the decision boundary between data clusters. Each hidden neuron has an RBF activation function that calculates the similarity between the input and a stored prototype in that neuron. Having more prototypes results in a more complex decision boundary, which means higher accuracy. However, it results of more computations to evaluate the network.

Figure 8.1 shows the structure of RBF network in comparison with the Multilayer Perceptron network. Inspecting this figure, it may be seen that the arrows between the input layer and the hidden neurons in the RBF network represent the Euclidean distance between the input vector and the prototypes stored in the hidden neurons. On the other side, in MLP, arrows between the input layer and output layer represent weights. Moreover, in RBF networks, activation functions in the hidden nodes are Gaussian basis functions while in MLP the sigmoidal functions are typically used.

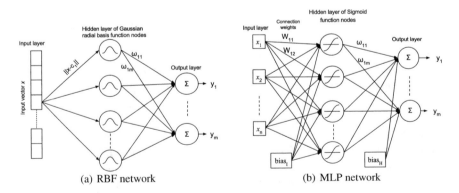

(a) RBF network   (b) MLP network

Fig. 8.1 Representation of RBF and MLP networks

RBF ANN process works as follows, first the input data enter the network through the input layer. After that, each neuron in the RBF layer (hidden layer) calculates the similarity between the input data and the prototype stored inside it, using the nonlinear Gaussian function shown in Eq. 8.1.

$$\phi(\|x - c_j\|) = exp - (\frac{\|x - c_j\|^2}{2\sigma_j^2})$$

(8.1)

where $\|x - c_j\|$ is Euclidean norm.

The output of the RBF is calculated using weighted average method by the following equation:

$$y_i = \sum_{j=1}^{n} \omega_{ji}\phi_j(x)$$

(8.2)

where ω_{ji} represents the ith weight between the hidden layer and output layer, and n represents the number of hidden nodes.

The output of the RBF neuron is closer to 1 whenever the similarity between the input and the prototype is high, and close to zero otherwise. The output layer neurons takes the weighted sum of every RBF neuron output in order to decide the class label. Which means that every RBF neuron contributes in the labeling decision, higher similarity has larger contribution.

8.2.1 Classical Radial Basis Function Network Training

Classical RBF process mainly depends on three points, the prototypes inside each RBF neurons and how to be chosen perfectly, the beta value in the similarity equation, and the weights between hidden layer and output layer (which affects the last decision). Choosing the prototypes can be done using many approaches such as choosing random data points from the data, or using K-Means clustering approach and use the clusters centers as prototypes or any other approach you may choose. Using K-Means clustering is the most used approach in the literature as it helps in smartly choosing small number of RBF neurons (K neurons), where each neuron represents a cluster in the data. Moreover, having only K neurons does not affect the complexity of the RBF network nor the accuracy of the classification decision.

Beta coefficient in the RBF activation function controls the width of the bell curve, and should be determined in a manner that optimises the fit between the activation function and the data. When K-Means is used to choose the RBF neurons prototypes then Beta can be set using the following Equation:

$$Beta = \frac{1}{2 \times \sigma^2}$$

(8.3)

where σ equals to the average distance between all points in the cluster and the cluster center.

Training the output weights is the third important parameter to set for RBF to work perfectly. Training these weights can be done using the Gradient decent which is an optimisation technique that takes the outputs of the RBF neurons as input and optimise the weights according to them. Gradient descent must be run separately for each output node. The following subsections describe more details about K-Means and gradient decent method that selected in this work as a classical approach for optimising the connection weights.

8.2.1.1 K-Means

K-Means is considered one of the most efficient clustering algorithms that used in many applications in the literature. K-Means clustering has many advantages, such as simplicity to implement and has good performance with large dataset. K-Means is a partitioning clustering algorithm, where the objective is to maximise the similarity between the members in each cluster, and minimise the similarity between the members in different clusters. The main idea of the K-Means clustering is to define k centers, one center for each cluster. The data points are assigned to the proper cluster based on the minimum distances to all cluster centers. After that, cluster centers should be modified in an iterative way by calculating the mean of cluster's members to achieve the best clustering quality (The squared error function). This process is continued until centers do not change any more.

8.2.1.2 Gradient Decent

Gradient Decent (GD) is considered an optimisation algorithm uses the first-order derivative calculation to find a local minimum of a function. The algorithm applies consecutive steps to find the gradient of the objective function at the current point. The output of a RBF network can be represented as shown in Eqs. 8.4 and 8.5, while the error function E is given in Eq. 8.7, where $\hat{y}_{i,k}$ is the response value of the ith output unit, and $y_{i,k}$ is the actual response. The Gradient decent algorithms can be used to find the solution matrix W as shown in Eq. 8.7, where η is a small decreasing value called the learning rate.

$$\hat{y} = (y_1, y_2, ..., y_m) = \begin{bmatrix} \omega_{11} & \omega_{11} & .. & \omega_{1m} \\ \omega_{21} & \omega_{21} & .. & \omega_{2m} \\ . & . & ... & . \\ . & . & ... & . \\ \omega_{l1} & \omega_{l1} & .. & \omega_{lm} \end{bmatrix} \begin{bmatrix} \phi_1(x) \\ \phi_2(x) \\ . \\ . \\ \phi_m(x) \end{bmatrix} \tag{8.4}$$

$$O = W \cdot H \tag{8.5}$$

$$E = \frac{1}{2} \sum_{k=1}^{M} \sum_{i=1}^{L} (y_{i,k} - \hat{y}_{i,k})^2 \tag{8.6}$$

$$\omega_{ij} = \omega_{ij} - \eta \frac{\partial E}{\partial \omega} \tag{8.7}$$

In this work, the Conjugate Gradient (CG) is used to optimise the weights in the standard RBF network. CG is a special type of gradient descent with regularization that used to compute search directions. The CG uses a line search with quadratic, and cubic polynomial approximations. The stopping criteria that used in CG is the Wolfe-Powell, and CG guesses the initial step sizes using slope ratio method.

8.3   Evolutionary Algorithms for Training Radial Basis Function Networks

In contrast to the classical approach where RBF network are trained in two independent phases, the proposed BBO-based approach [13] searches for all RBF network parameters simultaneously. The parameters are the centers, widths and connection weights including the bias terms. In the proposed training algorithm, each habitat is encoded to represent these parameters as shown in Fig. 8.2 where C_i is the center of the hidden neuron i, σ_i is the width of that neuron and ω_{ij} is weight connecting between neuron i and output unit j. Habitats are implemented as real vectors with a length D which can be calculated as follows: suppose that n is number of hidden neurons, I is the number of features in the dataset and m is the number of output units then D can be calculated as given in Eq. 8.8.

$$D = (n \times I) + n + (n \times m) + m \tag{8.8}$$

In order to evaluate the fitness value of the solutions (candidate RBF networks), the Mean Squared Error (MSE) is calculated over all training samples for each solution. MSE can be given as in Eq. 8.9 where y is the actual output, $\hat{y}$ is the estimated output and k is the total number of instances in the training dataset.

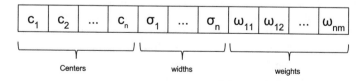

Fig. 8.2 Representation of BBO individuals structure

$$MSE = \frac{1}{k} \sum_{i=1}^{k} (y - \hat{y})^2 \qquad (8.9)$$

Based on the encoding scheme and the fitness evaluation described above, EA algorithms can train the RBF networks. To do so, an EA first creates a set of random candidate solutions, which includes RBF networks with random connection weights and biases. The algorithm then repeatedly calculates the MSE for all the RBF networks when classifying the training data. The MSE shows which "random" RBF is better. Based on the rules discussed above, the EA algorithm creates a set of new RBF networks considering the best RBF networks found so far. The process of calculating MSEs and improving the RBFs continues until the satisfaction of the end criterion, which could be a threshold or maximum iterations. It should be noted that the average MSE is calculated when classifying all training samples in the dataset for each RBF network in the proposed EA-based trainers. Therefore, the computational complexity is of $O(ntd)$ where n is the number of random RBF networks, t indicates the maximum iterations, and d shows the number of training samples in the dataset.

8.4   Experiments and Results

In this section, a set of EA training algorithms are evaluated on twelve data sets. A diverse range of EAs including conventional (GA, PSO, ACO, ES, PBIL, and DE) and recent algorithms (Firefly, Cuckoo Search, Artificial Bee Colony, and Bat Algorithm) are used. The algorithms are compared with the RBFclassic (Gradient-based) technique, which is considered the common method for training the RBF neural network.

8.4.1   Experimental Setup

The Matlab R2010b is used to implement the proposed BBO trainer and other algorithms. All datasets are divided using 66%, 34% for training and testing, respectively. Ten independent runs are conducted for all experiments, and 250 iterations is considered for each run. Moreover, the population size is fixed to 50 individuals for all algorithms. The parameter settings for each algorithm are shown in Table 8.1.

In CS, besides the population size, the discovery rate p_α is the only parameter needs to be tuned. p_α is set to 0.25 since it was stated in [14] that this value is sufficient for most optimisation problems. For Firefly, Beta is set to 1 as it was reported in [15] that parametric studies suggest setting the value of Beta to 1 can be used for most applications while gamma can be set to $1/\sqrt{L}$ where L is a scaling factor and if the scaling variations are not significant, then we can set gamma $= O(1)$. Alpha is roughly tuned and set to 0.2. Same values were used and applied in previous studies as in [16].

Table 8.1 The metaheuristic algorithms with initial parameters

Algorithm	Parameter	Value
GA	• Crossover probability	0.9
	• Mutation probability	0.1
	• Selection mechanism	Stochastic universal sampling
PSO	• Acceleration constants	[2.1, 2.1]
	• Inertia weights	[0.9, 0.6]
DE	• Crossover probability	0.9
	• Differential weight	0.5
BBO	• Habitat modification probability	1.0
	• Immigration probability	[0, 1]
	• Step size	1.0
	• Maximum immigration	1.0
	• Migration rates	1.0
	• Mutation probability	0.05
ACO	• Initial pheromone (τ)	$1e - 06$
	• Pheromone update constant (Q)	20
	• Pheromone constant (q)	1
	• Global pheromone decay rate (p_g)	0.9
	• Local pheromone decay rate (p_t)	0.5
	• Pheromone sensitivity (α)	1
	• Visibility sensitivity (β)	5
ES	• λ	10
	• σ	1
PBIL	• Learning rate	0.05
	• Good population member	1
	• Bad population member	0
	• Elitism parameter	1
	• Mutational probability	0.1
FireFly	• Alpha	0.2
	• Beta	1
	• Gamma	1
Cuckoo	• Discovery rate P_α	0.25
ABC	Acceleration coefficient upper bound	1
BAT	Loudness	0.5
	Pulse rate	0.5
	Frequency minimum	0
	Frequency maximum	1

For PSO, the inertia weight is decreased in the interval [0.9, 0.6]. For GA, the crossover probability is usually set to a much high rate while the mutation probability is set to a much low probability. With a rough tuning, the crossover and mutation probabilities are set to 0.9 and 0.1, respectively. For DE, the DE/rand/1/bin variant is applied with the crossover probability and differential weight equal to 0.9 and 0.5 as applied and recommended in [17].

For ACO, ES and PBIL, all parameters are set as used and applied in [18]. And for ABC and Bat the defaults parameters are used [19, 20].

For BBO, we used the same parameters as in [18, 21] habitat modification probability is set to 1, immigration probability bounds per gene = [0, 1], step size is set to 1, maximum immigration and migration rates for each island is set to 1, while the mutation probability is set 0.05 as in [18].

However, it is worth mentioning that finding the best parameters of these algorithms is considered as another optimisation problem by itself and it is known as Meta-Optimisation. Therefore fine tuning the optimisation algorithms is out of scope if this book. All dataset are normalized to the interval of [0, 1].

An RBF network with large number of neurons in the hidden layer may achieve good results based on the training data; however, this could lead to a bad generalization [22]. In our experiments, we assess the performance of the proposed training algorithm based on different number of neurons in the hidden layer: 4, 6, 8, 10 respectively.

In the experiments, we used five different measurements to evaluate the developed RBF network models. The measurements are Accuracy, Specificity, Sensitivity, Complexity and MSE. MSE was given previously in Eq. 8.9 while the rest are calculated using the following Eqs. 8.10, 8.11, 8.12 and 8.13, respectively. Accuracy, Specificity, Sensitivity and MSE assess the prediction accuracy while the complexity equation reflects the network structure complexity based on the number of neurons.

$$Accuracy = \frac{Number\ of\ correctly\ classified\ instances}{Total\ number\ of\ instances} \tag{8.10}$$

$$Specificity = \frac{Number\ of\ predicted\ instances\ of\ negative\ class}{Number\ of\ actual\ negative\ instances} \tag{8.11}$$

$$Sensitivity = \frac{Number\ of\ predicted\ instances\ of\ positive\ class}{Number\ of\ actual\ positive\ instances} \tag{8.12}$$

$$Complexity = \frac{1}{2}\sum_{i=1}^{|w|}(w_i)^2 \tag{8.13}$$

where $|w| = 2 * (n + 1)$, n is the number of neurons.

8.4.2 Datasets Description

The EA trainers are evaluated using twelve known real datasets, which are selected from UCI Repository [23]. All datasets contains two classes. Table 8.2 describes

Table 8.2 Summary of the classification data sets

Data set	#Features	#Training samples	#Testing samples	Accuracy of baseline classifier
Blood	4	493	255	0.7647
Breast	8	461	238	0.6597
Hepatitis	10	102	53	0.8113
Diabetes	8	506	262	0.6336
Vertebral	6	204	106	0.7075
Diagnosis I	6	79	41	0.5366
Diagnosis II	6	79	41	0.5366
Parkinson	22	128	67	0.7612
Liver	6	227	118	0.5763
Sonar	60	137	71	0.5352
German	24	660	340	0.6706
Australian	14	455	235	0.5702

these datasets in terms of number of features, training samples, testing samples and the accuracy of the baseline classifier for each dataset. The baseline classifier is the Zero Rule classifier or ZeroR for short. ZeroR is the simplest classifier which relies only on the output class by simply predicting the majority class.

8.4.3 Results

Table 8.3 shows the the results in term of the average accuracy (AVE) and standard deviation (STD), as well as the best accuracy result of the BBO and other algorithms on Blood dataset. The table reports the results with different number of neurons in the hidden layer. The best accuracy results are highlighted in bold. According to the results of AVE, STD, and best results using 4 neurons, BBO is able to classify 77.1% of the test samples, which is more than PSO, ACO, ES, PBIL, DE and ABC results and with slight difference with GA, FireFly, Cuckoo, BAT and RBFclassic. Furthermore, BBO outperforms all other methods using 6 neurons and 8 neurons with accuracy rates 77.29% and 77.45%, respectively. In addition, the accuracy results of RBFclassic, BAT and BBO using 10 neurons are very close for this dataset, and the three algorithms outperform the other methods.

The accuracy results of BBO and other optimisers for Breast cancer dataset are presented in Table 8.4. According to the results of AVE, STD, and best results, BBO outperforms all other methods using 4, 6 and 10 neurons. Moreover, the BBO able to classify 96.55%, 97.61%, 96.97%, and 97.86% of the test samples using 4, 6, 8, and 10 neurons, respectively.

Table 8.3 The average accuracy, and standard deviation results of the Blood dataset using different algorithms

Algorithm	The number of neurons of hidden layer							
	4		6		8		10	
	AVE ± STD	Best	AVE ± STD	Best	AVE ± STD	Best	AVE ± STD	Best
BBO	0.7710 ± 0.0046	0.7765	0.7729 ± 0.0039	0.7765	0.7745 ± 0.0028	0.7765	0.7733 ± 0.0036	0.7765
GA	0.7718 ± 0.0045	0.7765	0.7714 ± 0.0045	0.7765	0.7722 ± 0.0057	0.7765	0.7718 ± 0.0052	0.7765
PSO	0.7671 ± 0.0033	0.7725	0.7690 ± 0.0039	0.7765	0.7663 ± 0.0027	0.7725	0.7671 ± 0.0053	0.7765
ACO	0.7031 ± 0.1515	0.7686	0.6929 ± 0.1305	0.7686	0.7290 ± 0.0523	0.7765	0.7639 ± 0.0045	0.7686
ES	0.7643 ± 0.0050	0.7725	0.7549 ± 0.0332	0.7765	0.7455 ± 0.0445	0.7686	0.7408 ± 0.0554	0.7725
PBIL	0.7671 ± 0.0096	0.7804	0.7663 ± 0.0059	0.7765	0.7616 ± 0.0118	0.7725	0.7667 ± 0.0081	0.7765
DE	0.7655 ± 0.0017	0.7686	0.7624 ± 0.0083	0.7725	0.7553 ± 0.0252	0.7725	0.7663 ± 0.0033	0.7725
FireFly	0.7722 ± 0.0043	0.7804	0.7694 ± 0.0045	0.7765	0.7686 ± 0.0041	0.7765	0.7694 ± 0.0055	0.7765
Cuckoo	0.7718 ± 0.0041	0.7765	0.7702 ± 0.0046	0.7765	0.7714 ± 0.0042	0.7765	0.7686 ± 0.0037	0.7725
RBFclassic	0.7765 ± 0.0000	0.7765	0.7686 ± 0.0000	0.7686	0.7686 ± 0.0000	0.7686	0.8118 ± 0.0000	0.8118
ABC	0.7639 ± 0.0086	0.7725	0.7659 ± 0.0026	0.7725	0.7663 ± 0.0027	0.7725	0.7663 ± 0.0038	0.7725
BAT	0.7729 ± 0.0034	0.7765	0.7725 ± 0.0037	0.7765	0.7718 ± 0.0045	0.7765	0.7757 ± 0.0052	0.7804

Table 8.4 The average accuracy and standard deviation results of the Breast dataset using different algorithms

| Algorithm | The number of neurons of hidden layer | | | | | | | |
| | 4 | | 6 | | 8 | | 10 | |
	AVE ± STD	Best	AVE ± STD	Best	AVE ± STD	Best	AVE ± STD	Best
BBO	0.9655 ± 0.0119	0.9790	0.9761 ± 0.0053	0.9832	0.9697 ± 0.0108	0.9832	0.9786 ± 0.0037	0.9832
GA	0.9647 ± 0.0168	0.9790	0.9639 ± 0.0143	0.9790	0.9689 ± 0.0121	0.9832	0.9664 ± 0.0097	0.9790
PSO	0.9294 ± 0.0510	0.9790	0.9294 ± 0.0438	0.9622	0.9315 ± 0.0272	0.9622	0.9118 ± 0.0333	0.9538
ACO	0.6765 ± 0.1691	0.8151	0.6303 ± 0.2806	0.9034	0.6013 ± 0.2291	0.9622	0.6853 ± 0.2100	0.9328
ES	0.7025 ± 0.1942	0.8866	0.5887 ± 0.2510	0.9244	0.6605 ± 0.2319	0.9412	0.6441 ± 0.1497	0.8445
PBIL	0.9311 ± 0.0361	0.9706	0.9105 ± 0.0914	0.9706	0.8874 ± 0.0475	0.9496	0.8319 ± 0.1982	0.9706
DE	0.7895 ± 0.1094	0.9412	0.6929 ± 0.2318	0.8908	0.8626 ± 0.0917	0.9706	0.8546 ± 0.0736	0.9664
FireFly	0.9567 ± 0.0086	0.9664	0.9563 ± 0.0147	0.9832	0.9462 ± 0.0239	0.9706	0.9660 ± 0.0090	0.9790
Cuckoo	0.9639 ± 0.0077	0.9790	0.9567 ± 0.0116	0.9748	0.9643 ± 0.0103	0.9790	0.9525 ± 0.0131	0.9706
RBFclassic	0.9580 ± 0.0000	0.9580	0.9622 ± 0.0000	0.9622	0.9538 ± 0.0000	0.9538	0.9454 ± 0.0000	0.9454
ABC	0.9613 ± 0.0090	0.9706	0.9546 ± 0.0106	0.9748	0.9248 ± 0.0443	0.9748	0.9311 ± 0.0329	0.9790
BAT	0.9550 ± 0.0480	0.9832	0.9685 ± 0.0112	0.9790	0.9744 ± 0.0046	0.9790	0.9693 ± 0.0136	0.9790

Table 8.5 presents the accuracy results of the Habitit dataset. The results of BBO are significantly better than all the other algorithms using 6 neurons. Moreover, BBO has better results compared with most algorithms using 4, 8, and 10 neurons as well.

The accuracy results of BBO and other training algorithms on Diabetes and Vertebral datasets are presented in Tables 8.6 and 8.7, respectively. According to the results of AVE, STD, and best results using 4, 6, 8, and 10 neurons, BBO outperforms most of other methods except RBFclassic which has better accuracy. These results support the merits of the BBO algorithm in training RBF networks.

The accuracy results of BBO and other algorithms on Diagnosis I and Diagnosis II datasets are presented in Tables 8.8 and 8.9, respectively. According to the results of AVE, STD, and best results, the results of BBO on the Diagnosis I are significantly better than all the other algorithms using different number of neurons. Furthermore, the BBO results on the Diagnosis II outperforms most of other optimises with 100%, 98.29%, and 99.51% using 4, 6, 8, 10 neurons, respectively. These results support the strength of the BBO algorithm in training RBF networks.

Tables 8.10 and 8.11 show the accuracy results of BBO and other algorithms on Parkinsons and Liver datasets, respectively. The accuracy results of BBO on the Parkinson, and Liver datasets outperform all other algorithms using different number of neurons.

The accuracy results of BBO and other training algorithms on Sonar and German datasets are presented in Tables 8.12 and 8.13, respectively. According to both dataset results using 4, 6, 8, and 10 neurons, BBO comes second after the RBFclassic method and it outperforms all other metaheuristic methods.

The accuracy results of BBO and other optimisers for Australian dataset are presented in Table 8.14. According to the results of using 4, 6, 8, and 10 neurons, BBO has superior classification accuracy results compared with other methods. Moreover, the BBO was able to classify 85.32%, 84.98%, 84.51%, and 85.11% of the test samples using 4, 6, 8, and 10 neurons, respectively.

To give a better insight on the classification performance regarding each class label, the specificity and sensitivity are measured and listed in Tables 8.15, 8.16, 8.17 and 8.18 for RBF networks with 2, 6, 8 and 10 neurons in the hidden layer respectively. According to these tables, it can be noticed that RBF networks optimised by BBO has higher and more balanced specificity and sensitivity than most of the other optimisers in the following datasets: Breast cancer, Habitit, Vertebral, Diagnosis I, Diagnosis II, Parkinsons, Liver, Sonar and Australian.

To summarize the above results, we can note that BBO outperform most of other algorithms, which supports the merits of the BBO algorithm in training MLPs. Moreover, and to support this summary, Friedman statistical test is calculated to check the significance of the accuracy results. Friedman test is accomplished by ranking the different trainers (BBO, GA, PSO, ACO, ES, PBIL, DE, FireFly, Cuckoo, ABC, BAT and RBFclassic) based on the average accuracy values for each dataset using different neurons. Table 8.19 shows the average ranks for each technique in using Friedman test for 4, 6, 8, and 10 neurons. The Friedman test in the Table 8.19 shows that significant difference exist between the 12 trainers (lower is better). In term of F-test ranking, BBO outperforms other trainers for all number of neurons that used.

Table 8.5 The average accuracy and standard deviation results of the Habitit dataset using different algorithms

Algorithm	The number of neurons of hidden layer							
	4		6		8		10	
	AVE ± STD	Best	AVE ± STD	Best	AVE ± STD	Best	AVE ± STD	Best
BBO	0.8377 ± 0.0221	0.8868	0.8472 ± 0.0165	0.8679	0.8283 ± 0.0188	0.8679	0.8302 ± 0.0126	0.8491
GA	0.8453 ± 0.0232	0.8868	0.8321 ± 0.0165	0.8491	0.8453 ± 0.0318	0.9057	0.8472 ± 0.0259	0.9057
PSO	0.8358 ± 0.0252	0.8868	0.8245 ± 0.0296	0.8868	0.8321 ± 0.0273	0.8868	0.8245 ± 0.0268	0.8491
ACO	0.5472 ± 0.2871	0.8302	0.6642 ± 0.2200	0.9057	0.6585 ± 0.1972	0.8491	0.5528 ± 0.2655	0.8302
ES	0.7472 ± 0.1249	0.8113	0.5830 ± 0.2892	0.8113	0.6887 ± 0.2292	0.8491	0.7057 ± 0.1845	0.8679
PBIL	0.8283 ± 0.0259	0.8679	0.8264 ± 0.0214	0.8679	0.8321 ± 0.0259	0.8679	0.8189 ± 0.0419	0.9245
DE	0.8000 ± 0.0497	0.8302	0.7038 ± 0.1797	0.8491	0.6755 ± 0.2328	0.8302	0.7887 ± 0.0716	0.8491
FireFly	0.8547 ± 0.0252	0.9057	0.8321 ± 0.0165	0.8491	0.8340 ± 0.0292	0.8679	0.8340 ± 0.0149	0.8491
Cuckoo	0.8623 ± 0.0282	0.8868	0.8396 ± 0.0160	0.8679	0.8472 ± 0.0243	0.8868	0.8245 ± 0.0296	0.8679
RBFclassic	0.8302 ± 0.0000	0.8302	0.8302 ± 0.0000	0.8302	0.8113 ± 0.0000	0.8113	0.7736 ± 0.0000	0.7736
ABC	0.8340 ± 0.0306	0.8868	0.8358 ± 0.0252	0.8679	0.8321 ± 0.0273	0.8868	0.8377 ± 0.0347	0.8868
BAT	0.8509 ± 0.0165	0.8868	0.8453 ± 0.0149	0.8679	0.8321 ± 0.0208	0.8679	0.8264 ± 0.0195	0.8679

Table 8.6 The average accuracy and standard deviation results of the Diabetes dataset using different algorithms

Algorithm	The number of neurons of hidden layer							
	4		6		8		10	
	AVE ± STD	Best	AVE ± STD	Best	AVE ± STD	Best	AVE ± STD	Best
BBO	0.6950 ± 0.0237	0.7443	0.6996 ± 0.0268	0.7481	0.6996 ± 0.0181	0.7214	0.7160 ± 0.0190	0.7481
GA	0.6702 ± 0.0223	0.6947	0.6863 ± 0.0239	0.7176	0.6916 ± 0.0275	0.7405	0.7000 ± 0.0206	0.7290
PSO	0.6599 ± 0.0360	0.7061	0.6645 ± 0.0310	0.7176	0.6752 ± 0.0291	0.7023	0.6569 ± 0.0257	0.7023
ACO	0.6008 ± 0.0855	0.6527	0.5767 ± 0.0921	0.6489	0.5687 ± 0.1170	0.6489	0.5668 ± 0.1248	0.6527
ES	0.5794 ± 0.1263	0.6794	0.6244 ± 0.0510	0.7176	0.5679 ± 0.0880	0.6489	0.6130 ± 0.0452	0.6603
PBIL	0.6676 ± 0.0242	0.7176	0.6630 ± 0.0363	0.7557	0.6626 ± 0.0425	0.7557	0.6401 ± 0.0282	0.6794
DE	0.6378 ± 0.0120	0.6527	0.6176 ± 0.0642	0.7023	0.6263 ± 0.0686	0.7099	0.6126 ± 0.0743	0.6679
FireFly	0.6771 ± 0.0421	0.7595	0.6931 ± 0.0374	0.7634	0.6863 ± 0.0365	0.7557	0.6763 ± 0.0318	0.7290
Cuckoo	0.6817 ± 0.0375	0.7710	0.6863 ± 0.0223	0.7405	0.6954 ± 0.0267	0.7443	0.6672 ± 0.0275	0.7099
RBFclassic	0.7405 ± 0.0000	0.7405	0.7672 ± 0.0000	0.7672	0.7748 ± 0.0000	0.7748	0.7481 ± 0.0000	0.7481
ABC	0.6985 ± 0.0250	0.7405	0.6989 ± 0.0219	0.7481	0.6985 ± 0.0305	0.7519	0.6782 ± 0.0282	0.7214
BAT	0.7034 ± 0.0228	0.7366	0.7118 ± 0.0176	0.7405	0.7069 ± 0.0239	0.7443	0.7088 ± 0.0254	0.7634

Table 8.7 The average accuracy and standard deviation results of the Vertebral dataset using different algorithms

Algorithm	The number of neurons of hidden layer							
	4		6		8		10	
	AVE ± STD	Best	AVE ± STD	Best	AVE ± STD	Best	AVE ± STD	Best
BBO	0.7632 ± 0.0340	0.8019	0.7708 ± 0.0298	0.8208	0.7840 ± 0.0201	0.8208	0.7792 ± 0.0260	0.8302
GA	0.7604 ± 0.0236	0.8019	0.7519 ± 0.0281	0.8019	0.7623 ± 0.0217	0.8113	0.7708 ± 0.0236	0.8113
PSO	0.7406 ± 0.0425	0.8302	0.7104 ± 0.0256	0.7453	0.7255 ± 0.0525	0.8019	0.7283 ± 0.0444	0.8113
ACO	0.6189 ± 0.1727	0.7075	0.6453 ± 0.1328	0.7547	0.6217 ± 0.1547	0.7830	0.6094 ± 0.1458	0.7075
ES	0.7208 ± 0.0549	0.8302	0.6698 ± 0.0712	0.7736	0.6481 ± 0.1059	0.7358	0.6613 ± 0.0805	0.7264
PBIL	0.7368 ± 0.0421	0.8396	0.7406 ± 0.0339	0.7925	0.7113 ± 0.0408	0.7925	0.7000 ± 0.0307	0.7453
DE	0.6943 ± 0.0214	0.7264	0.7019 ± 0.0244	0.7453	0.6698 ± 0.0868	0.7358	0.6623 ± 0.1048	0.7547
FireFly	0.7443 ± 0.0365	0.7925	0.7519 ± 0.0263	0.7925	0.7519 ± 0.0248	0.7925	0.7575 ± 0.0336	0.8113
Cuckoo	0.7557 ± 0.0296	0.8019	0.7500 ± 0.0223	0.7736	0.7462 ± 0.0296	0.8113	0.7566 ± 0.0481	0.8396
RBFclassic	0.8208 ± 0.0000	0.8208	0.8302 ± 0.0000	0.8302	0.8302 ± 0.0000	0.8302	0.8302 ± 0.0000	0.8302
ABC	0.7208 ± 0.0271	0.7642	0.7264 ± 0.0537	0.8302	0.7047 ± 0.0199	0.7358	0.7019 ± 0.0214	0.7264
BAT	0.7594 ± 0.0463	0.8208	0.7632 ± 0.0245	0.8113	0.7623 ± 0.0217	0.8113	0.7783 ± 0.0253	0.8113

Table 8.8 The average accuracy and standard deviation results of the Diagnosis I dataset using different algorithms

Algorithm	The number of neurons of hidden layer							
	4		6		8		10	
	AVE ± STD	Best	AVE ± STD	Best	AVE ± STD	Best	AVE ± STD	Best
BBO	1.0000 ± 0.0000	1.0000	1.0000 ± 0.0000	1.0000	1.0000 ± 0.0000	1.0000	0.9780 ± 0.0694	1.0000
GA	0.9951 ± 0.0154	1.0000	1.0000 ± 0.0000	1.0000	1.0000 ± 0.0000	1.0000	0.9707 ± 0.0926	1.0000
PSO	0.8634 ± 0.0899	0.9512	0.8171 ± 0.1357	1.0000	0.7610 ± 0.1114	0.9756	0.8439 ± 0.0672	0.9512
ACO	0.5293 ± 0.1436	0.7561	0.5732 ± 0.1869	0.9268	0.5244 ± 0.2179	0.9756	0.5195 ± 0.1048	0.7805
ES	0.5244 ± 0.1068	0.7561	0.5634 ± 0.0533	0.6341	0.6024 ± 0.1602	0.8780	0.5634 ± 0.0817	0.6829
PBIL	0.8146 ± 0.1455	1.0000	0.8220 ± 0.1260	1.0000	0.7073 ± 0.1988	0.9268	0.8000 ± 0.1739	1.0000
DE	0.6341 ± 0.1751	0.8780	0.7024 ± 0.1476	0.8780	0.5951 ± 0.1771	1.0000	0.6317 ± 0.2177	0.9024
FireFly	0.9805 ± 0.0427	1.0000	0.9659 ± 0.0504	1.0000	0.9463 ± 0.0561	1.0000	0.9683 ± 0.0431	1.0000
Cuckoo	0.9707 ± 0.0926	1.0000	0.9805 ± 0.0427	1.0000	0.9878 ± 0.0263	1.0000	0.9683 ± 0.0920	1.0000
RBFclassic	0.8780 ± 0.0000	0.8780	0.8780 ± 0.0000	0.8780	1.0000 ± 0.0000	1.0000	0.9756 ± 0.0000	0.9756
ABC	0.9244 ± 0.0825	1.0000	0.7659 ± 0.1207	0.9756	0.7951 ± 0.0997	0.9268	0.8683 ± 0.0997	1.0000
BAT	0.9878 ± 0.0386	1.0000	0.9780 ± 0.0694	1.0000	1.0000 ± 0.0000	1.0000	1.0000 ± 0.0000	1.0000

Table 8.9 The average accuracy and standard deviation results of the Diagnosis II dataset using different algorithms

Algorithm	The number of neurons of hidden layer							
	4		6		8		10	
	AVE ± STD	Best	AVE ± STD	Best	AVE ± STD	Best	AVE ± STD	Best
BBO	1.0000 ± 0.0000	1.0000	0.9829 ± 0.0364	1.0000	0.9951 ± 0.0154	1.0000	0.9927 ± 0.0231	1.0000
GA	0.9976 ± 0.0077	1.0000	1.0000 ± 0.0000	1.0000	1.0000 ± 0.0000	1.0000	1.0000 ± 0.0000	1.0000
PSO	0.8220 ± 0.0948	1.0000	0.8390 ± 0.1302	1.0000	0.8854 ± 0.0830	1.0000	0.8537 ± 0.1222	1.0000
ACO	0.5732 ± 0.1086	0.7073	0.5488 ± 0.1317	0.8537	0.5195 ± 0.0976	0.7317	0.5683 ± 0.2136	0.9268
ES	0.4780 ± 0.1036	0.6098	0.5902 ± 0.1412	0.8049	0.6220 ± 0.2173	1.0000	0.6561 ± 0.1005	0.8049
PBIL	0.8512 ± 0.1779	1.0000	0.8073 ± 0.1169	1.0000	0.7512 ± 0.1957	1.0000	0.8098 ± 0.0967	1.0000
DE	0.6122 ± 0.1303	0.8049	0.6268 ± 0.1965	1.0000	0.6317 ± 0.1652	0.8537	0.6707 ± 0.1157	0.8293
FireFly	0.9707 ± 0.0524	1.0000	0.9585 ± 0.0502	1.0000	0.9512 ± 0.0660	1.0000	0.9512 ± 0.0586	1.0000
Cuckoo	1.0000 ± 0.0000	1.0000	0.9756 ± 0.0325	1.0000	0.9659 ± 0.0611	1.0000	0.9463 ± 0.0648	1.0000
RBFclassic	1.0000 ± 0.0000	1.0000	1.0000 ± 0.0000	1.0000	1.0000 ± 0.0000	1.0000	1.0000 ± 0.0000	1.0000
ABC	0.9537 ± 0.0775	1.0000	0.9293 ± 0.0848	1.0000	0.8951 ± 0.0927	1.0000	0.8756 ± 0.0841	1.0000
BAT	0.9927 ± 0.0165	1.0000	0.9634 ± 0.0773	1.0000	0.9927 ± 0.0231	1.0000	1.0000 ± 0.0000	1.0000

Table 8.10 The average accuracy and standard deviation results of the Parkinson dataset using different algorithms

Algorithm	The number of neurons of hidden layer							
	4		6		8		10	
	AVE ± STD	Best	AVE ± STD	Best	AVE ± STD	Best	AVE ± STD	Best
BBO	0.8358 ± 0.0416	0.8657	0.8493 ± 0.0164	0.8657	0.8522 ± 0.0110	0.8657	0.8642 ± 0.0192	0.8955
GA	0.8194 ± 0.0458	0.8955	0.8194 ± 0.0431	0.8657	0.8373 ± 0.0192	0.8657	0.8478 ± 0.0231	0.8806
PSO	0.7716 ± 0.0483	0.8657	0.7806 ± 0.0559	0.8657	0.7746 ± 0.0495	0.8657	0.7478 ± 0.1213	0.8657
ACO	0.6090 ± 0.2162	0.8060	0.6522 ± 0.2121	0.8209	0.5866 ± 0.2497	0.8507	0.5821 ± 0.2338	0.7910
ES	0.5672 ± 0.2055	0.7761	0.5104 ± 0.2386	0.7612	0.5090 ± 0.2293	0.7761	0.5343 ± 0.2351	0.7761
PBIL	0.7776 ± 0.0164	0.8060	0.7716 ± 0.0173	0.8060	0.7612 ± 0.0000	0.7612	0.7612 ± 0.0000	0.7612
DE	0.5478 ± 0.2429	0.7761	0.5552 ± 0.2491	0.7612	0.7403 ± 0.0672	0.7910	0.5463 ± 0.2372	0.7761
FireFly	0.7910 ± 0.0244	0.8358	0.7881 ± 0.0271	0.8209	0.7955 ± 0.0338	0.8507	0.7955 ± 0.0467	0.8806
Cuckoo	0.8239 ± 0.0297	0.8507	0.8134 ± 0.0354	0.8507	0.8030 ± 0.0252	0.8358	0.7896 ± 0.0277	0.8209
RBFclassic	0.7910 ± 0.0000	0.7910	0.8209 ± 0.0000	0.8209	0.8358 ± 0.0000	0.8358	0.8507 ± 0.0000	0.8507
ABC	0.7955 ± 0.0330	0.8358	0.7985 ± 0.0292	0.8358	0.7791 ± 0.0321	0.8507	0.7642 ± 0.0609	0.8358
BAT	0.8254 ± 0.0273	0.8657	0.8463 ± 0.0299	0.8955	0.8149 ± 0.0399	0.8806	0.8343 ± 0.0192	0.8507

Table 8.11 The average accuracy and standard deviation results of the Liver dataset using different algorithms

Algorithm	The number of neurons of hidden layer							
	4		6		8		10	
	AVE ± STD	Best	AVE ± STD	Best	AVE ± STD	Best	AVE ± STD	Best
BBO	0.6288 ± 0.0298	0.6780	0.6500 ± 0.0256	0.6864	0.6568 ± 0.0266	0.6949	0.6915 ± 0.0436	0.7542
GA	0.6059 ± 0.0358	0.6864	0.6068 ± 0.0260	0.6610	0.6466 ± 0.0277	0.6949	0.6288 ± 0.0379	0.6864
PSO	0.5797 ± 0.0256	0.6271	0.5805 ± 0.0517	0.7034	0.5898 ± 0.0260	0.6271	0.5746 ± 0.0450	0.6356
ACO	0.5136 ± 0.0601	0.5763	0.5068 ± 0.0675	0.5763	0.5203 ± 0.0655	0.5932	0.5220 ± 0.0634	0.5847
ES	0.5508 ± 0.0504	0.5763	0.5051 ± 0.0663	0.5763	0.5339 ± 0.0744	0.5847	0.5339 ± 0.0594	0.6102
PBIL	0.5627 ± 0.0582	0.6356	0.5720 ± 0.0286	0.6271	0.5593 ± 0.0325	0.5932	0.5542 ± 0.0272	0.5847
DE	0.5610 ± 0.0232	0.5932	0.5525 ± 0.0527	0.6186	0.5432 ± 0.0612	0.6102	0.5220 ± 0.0638	0.6356
FireFly	0.6169 ± 0.0236	0.6525	0.6051 ± 0.0398	0.6864	0.5856 ± 0.0220	0.6271	0.6110 ± 0.0431	0.6780
Cuckoo	0.6229 ± 0.0302	0.6695	0.6102 ± 0.0496	0.6864	0.6178 ± 0.0380	0.6780	0.5983 ± 0.0335	0.6356
RBFclassic	0.5763 ± 0.0000	0.5763	0.5932 ± 0.0000	0.5932	0.6017 ± 0.0000	0.6017	0.6102 ± 0.0000	0.6102
ABC	0.5712 ± 0.0223	0.6186	0.5653 ± 0.0400	0.6102	0.5720 ± 0.0358	0.6186	0.5534 ± 0.0383	0.5932
BAT	0.6449 ± 0.0287	0.6949	0.6254 ± 0.0312	0.6695	0.6568 ± 0.0166	0.6864	0.6483 ± 0.0332	0.6949

Table 8.12 The average accuracy and standard deviation results of the Sonar dataset using different algorithms

Algorithm	The number of neurons of hidden layer							
	4		6		8		10	
	AVE ± STD	Best	AVE ± STD	Best	AVE ± STD	Best	AVE ± STD	Best
BBO	0.7549 ± 0.0365	0.8028	0.7296 ± 0.0608	0.8169	0.7197 ± 0.0698	0.8028	0.7380 ± 0.0431	0.7887
GA	0.6507 ± 0.1148	0.7746	0.5718 ± 0.1016	0.7183	0.5254 ± 0.1208	0.7746	0.6183 ± 0.1166	0.7324
PSO	0.5352 ± 0.0594	0.6197	0.4944 ± 0.0438	0.5775	0.4887 ± 0.0953	0.6761	0.5000 ± 0.0761	0.6197
ACO	0.5028 ± 0.0567	0.6338	0.5014 ± 0.0451	0.5915	0.5296 ± 0.0853	0.6901	0.5113 ± 0.0358	0.5352
ES	0.4930 ± 0.0398	0.5493	0.5099 ± 0.0759	0.7042	0.5014 ± 0.0498	0.5915	0.4915 ± 0.0490	0.5775
PBIL	0.4648 ± 0.0000	0.4648	0.4648 ± 0.0000	0.4648	0.4648 ± 0.0000	0.4648	0.4648 ± 0.0000	0.4648
DE	0.5085 ± 0.0348	0.5352	0.5239 ± 0.0459	0.5915	0.4944 ± 0.0462	0.5493	0.5000 ± 0.0840	0.6479
FireFly	0.5408 ± 0.0876	0.7042	0.5254 ± 0.0903	0.6338	0.5155 ± 0.1143	0.7183	0.5056 ± 0.0672	0.6338
Cuckoo	0.5423 ± 0.0772	0.7042	0.5423 ± 0.0851	0.6479	0.5549 ± 0.1012	0.7183	0.5268 ± 0.0930	0.6901
RBFclassic	0.8451 ± 0.0000	0.8451	0.7746 ± 0.0000	0.7746	0.7746 ± 0.0000	0.7746	0.8028 ± 0.0000	0.8028
ABC	0.5183 ± 0.0555	0.6197	0.5225 ± 0.0764	0.7183	0.5465 ± 0.0875	0.6761	0.4775 ± 0.0776	0.6197
BAT	0.5577 ± 0.1152	0.6761	0.5634 ± 0.0888	0.6761	0.6014 ± 0.1255	0.7324	0.5901 ± 0.1162	0.8028

Table 8.13 The average accuracy and standard deviation results of the German dataset using different algorithms

Algorithm	The number of neurons of hidden layer							
	4		6		8		10	
	AVE ± STD	Best	AVE ± STD	Best	AVE ± STD	Best	AVE ± STD	Best
BBO	0.7091 ± 0.0268	0.7382	0.7094 ± 0.0193	0.7382	0.7138 ± 0.0150	0.7441	0.7191 ± 0.0098	0.7353
GA	0.6756 ± 0.0059	0.6882	0.6809 ± 0.0222	0.7147	0.6779 ± 0.0191	0.7029	0.6794 ± 0.0211	0.7059
PSO	0.5638 ± 0.1095	0.6765	0.5532 ± 0.0947	0.6765	0.5556 ± 0.1034	0.6765	0.5247 ± 0.1112	0.6706
ACO	0.4609 ± 0.1400	0.6706	0.5200 ± 0.1604	0.6706	0.5591 ± 0.1452	0.7000	0.4679 ± 0.1363	0.6588
ES	0.5685 ± 0.1381	0.6706	0.4747 ± 0.1662	0.6706	0.5156 ± 0.1552	0.6706	0.5615 ± 0.1166	0.6706
PBIL	0.6703 ± 0.0009	0.6706	0.6709 ± 0.0009	0.6735	0.6706 ± 0.0014	0.6735	0.6703 ± 0.0009	0.6706
DE	0.4406 ± 0.1436	0.6647	0.4824 ± 0.1635	0.6706	0.5265 ± 0.1435	0.6706	0.4862 ± 0.1395	0.6706
FireFly	0.5703 ± 0.1100	0.6706	0.5691 ± 0.1279	0.6912	0.6012 ± 0.1159	0.6765	0.6129 ± 0.0971	0.6882
Cuckoo	0.6682 ± 0.0048	0.6735	0.6579 ± 0.0168	0.6706	0.6429 ± 0.0435	0.6794	0.6471 ± 0.0392	0.6765
RBFclassic	0.7235 ± 0.0000	0.7235	0.7235 ± 0.0000	0.7235	0.7265 ± 0.0000	0.7265	0.7147 ± 0.0000	0.7147
ABC	0.6712 ± 0.0041	0.6824	0.6615 ± 0.0117	0.6824	0.6571 ± 0.0243	0.6765	0.6550 ± 0.0313	0.6971
BAT	0.6674 ± 0.0170	0.6941	0.6685 ± 0.0086	0.6824	0.6609 ± 0.0414	0.7059	0.6726 ± 0.0113	0.6912

Table 8.14 The average accuracy and standard deviation results of the Australian dataset using different algorithms

Algorithm	The number of neurons of hidden layer							
	4		6		8		10	
	AVE ± STD	Best	AVE ± STD	Best	AVE ± STD	Best	AVE ± STD	Best
BBO	0.8532 ± 0.0177	0.8894	0.8498 ± 0.0164	0.8638	0.8451 ± 0.0095	0.8638	0.8511 ± 0.0083	0.8638
GA	0.8413 ± 0.0149	0.8596	0.8255 ± 0.0175	0.8426	0.8362 ± 0.0234	0.8681	0.8455 ± 0.0117	0.8638
PSO	0.7157 ± 0.0666	0.8043	0.6630 ± 0.0533	0.7617	0.6209 ± 0.0847	0.7319	0.6272 ± 0.1146	0.7872
ACO	0.5098 ± 0.0998	0.6596	0.5885 ± 0.1160	0.8383	0.4962 ± 0.1538	0.7489	0.5289 ± 0.0935	0.6766
ES	0.5409 ± 0.0903	0.7064	0.5604 ± 0.1344	0.7362	0.5911 ± 0.0804	0.7362	0.5745 ± 0.1425	0.7319
PBIL	0.6885 ± 0.1087	0.8340	0.6949 ± 0.1498	0.8298	0.7085 ± 0.0926	0.7745	0.6009 ± 0.1197	0.7660
DE	0.6064 ± 0.0887	0.7106	0.5221 ± 0.1139	0.6809	0.5740 ± 0.0737	0.7064	0.6038 ± 0.1160	0.7404
FireFly	0.7574 ± 0.0721	0.8255	0.7694 ± 0.0606	0.8170	0.7400 ± 0.0507	0.8128	0.7421 ± 0.0784	0.8468
Cuckoo	0.7843 ± 0.0440	0.8298	0.7945 ± 0.0366	0.8340	0.7872 ± 0.0160	0.8128	0.7851 ± 0.0272	0.8085
RBFclassic	0.8511 ± 0.0000	0.8511	0.8468 ± 0.0000	0.8468	0.8426 ± 0.0000	0.8426	0.8426 ± 0.0000	0.8426
ABC	0.8064 ± 0.0235	0.8340	0.7574 ± 0.0569	0.8426	0.7489 ± 0.0530	0.8340	0.7694 ± 0.0547	0.8426
BAT	0.8281 ± 0.0405	0.8723	0.8340 ± 0.0261	0.8553	0.8362 ± 0.0200	0.8553	0.8255 ± 0.0379	0.8596

Table 8.15 The average sensitivity and specificity results of all datasets using different algorithms with (4 Neurons)

Dataset	BBO Sen.	BBO Spec.	GA Sen.	GA Spec.	PSO Sen.	PSO Spec.	ACO Sen.	ACO Spec.	ES Sen.	ES Spec.	PBIL Sen.	PBIL Spec.
Blood	1.0000	0.0267	0.9969	0.0400	0.9995	0.0117	0.8764	0.1400	0.9908	0.0283	0.9887	0.0467
Breast	0.9771	0.9432	0.9771	0.9407	0.9573	0.8753	0.7924	0.4519	0.8408	0.4346	0.9885	0.8198
Diabetes	0.3583	0.8898	0.2115	0.9355	0.2635	0.8892	0.2271	0.8169	0.2719	0.7572	0.1677	0.9566
Habitit	0.5500	0.9047	0.5300	0.9186	0.2700	0.9674	0.2900	0.6070	0.2700	0.8581	0.2700	0.9581
Vertebral	0.9027	0.4258	0.8933	0.4387	0.9120	0.3258	0.7587	0.2806	0.8893	0.3129	0.9133	0.3097
Diagnosis I	1.0000	1.0000	1.0000	0.9909	0.8684	0.8591	0.5526	0.5091	0.8632	0.2318	0.9684	0.6818
Diagnosis II	1.0000	1.0000	1.0000	0.9947	0.9136	0.7158	0.6273	0.5105	0.6000	0.3368	0.9091	0.7842
Parkinsons	0.9961	0.3250	0.9941	0.2625	0.9451	0.2188	0.6667	0.4250	0.5843	0.5125	0.9882	0.1063
Liver	0.1900	0.9515	0.0980	0.9794	0.1900	0.8662	0.3740	0.6162	0.1820	0.8221	0.2880	0.7647
Sonar	0.6447	0.8818	0.5053	0.8182	0.5895	0.4727	0.4342	0.5818	0.3184	0.6939	0.0000	1.0000
German	0.9513	0.2161	0.9820	0.0518	0.6811	0.3250	0.3833	0.6188	0.6925	0.3161	0.9996	0.0000
Australian	0.8426	0.8612	0.8485	0.8358	0.6614	0.7567	0.3535	0.6276	0.3901	0.6545	0.3317	0.9575

Dataset	DE Sen.	DE Spec.	FireFly Sen.	FireFly Spec.	Cuckoo Sen.	Cuckoo Spec.	RBFclassic Sen.	RBFclassic Spec.	ABC Sen.	ABC Spec.	BAT Sen.	BAT Spec.
Blood	0.9995	0.0050	0.9995	0.0333	1.0000	0.0300	0.9897	0.0833	0.9938	0.0167	0.9964	0.0467
Breast	0.9191	0.5383	0.9605	0.9494	0.9745	0.9432	0.9809	0.9136	0.9643	0.9556	0.9752	0.9160
Diabetes	0.0531	0.9759	0.2823	0.9054	0.3271	0.8867	0.5625	0.8434	0.3729	0.8867	0.4094	0.8735
Habitit	0.0600	0.9721	0.5300	0.9302	0.5000	0.9465	0.7000	0.8605	0.4100	0.9326	0.6000	0.9093
Vertebral	0.9387	0.1032	0.8587	0.4677	0.8920	0.4258	0.8667	0.7097	0.8427	0.4258	0.8720	0.4871
Diagnosis I	0.6842	0.5909	1.0000	0.9636	1.0000	0.9455	1.0000	0.7727	0.9526	0.9000	1.0000	0.9773
Diagnosis II	0.6273	0.5947	1.0000	0.9368	1.0000	1.0000	1.0000	1.0000	0.9909	0.9105	1.0000	0.9842
Parkinsons	0.5843	0.4313	0.9824	0.1812	0.9804	0.3250	0.9412	0.3125	0.9667	0.2500	0.9863	0.3125
Liver	0.1540	0.8603	0.2260	0.9044	0.2160	0.9221	0.5600	0.5882	0.2260	0.8250	0.3140	0.8882
Sonar	0.6553	0.3394	0.6237	0.4455	0.3105	0.8091	0.7895	0.9091	0.3816	0.6758	0.4158	0.7212
German	0.3224	0.6813	0.6618	0.3839	0.9864	0.0205	0.8947	0.3750	0.9982	0.0054	0.9807	0.0295
Australian	0.4208	0.7463	0.6881	0.8097	0.7396	0.8179	0.9010	0.8134	0.8069	0.8060	0.8069	0.8440

Table 8.16 The average sensitivity and specificity results of all datasets using different algorithms with (6 Neurons)

Dataset	BBO Sen.	BBO Spec.	GA Sen.	GA Spec.	PSO Sen.	PSO Spec.	ACO Sen.	ACO Spec.	ES Sen.	ES Spec.	PBIL Sen.	PBIL Spec.
Blood	0.9969	0.0450	1.0000	0.0283	0.9969	0.0283	0.8574	0.1583	0.9626	0.0800	0.9872	0.0483
Breast	0.9796	0.9691	0.9822	0.9284	0.9662	0.8580	0.6904	0.5136	0.6987	0.3753	0.9879	0.7605
Diabetes	0.3719	0.8892	0.3187	0.8988	0.2083	0.9283	0.2729	0.7524	0.1427	0.9030	0.1552	0.9566
Habitit	0.5900	0.9070	0.4900	0.9116	0.1300	0.9860	0.4900	0.7047	0.2500	0.6605	0.1600	0.9814
Vertebral	0.8787	0.5097	0.8880	0.4226	0.8893	0.2774	0.8253	0.2097	0.7413	0.4968	0.9253	0.2935
Diagnosis I	1.0000	1.0000	1.0000	1.0000	0.9053	0.7409	0.7263	0.4409	0.6211	0.5136	0.9789	0.6864
Diagnosis II	1.0000	0.9632	1.0000	1.0000	0.9182	0.7474	0.6182	0.4684	0.7136	0.4474	0.8727	0.7316
Parkinsons	0.9941	0.3875	1.0000	0.2437	0.9490	0.2437	0.7745	0.2625	0.5569	0.3625	0.9980	0.0500
Liver	0.3040	0.9044	0.1460	0.9456	0.3140	0.7765	0.4800	0.5265	0.3900	0.5897	0.2220	0.8294
Sonar	0.6579	0.8121	0.3447	0.8333	0.3289	0.6848	0.3974	0.6212	0.3605	0.6818	0.0000	1.0000
German	0.9500	0.2196	0.9311	0.1714	0.6478	0.3607	0.5561	0.4464	0.4360	0.5536	0.9996	0.0018
Australian	0.8505	0.8493	0.8178	0.8313	0.4941	0.7903	0.4238	0.7127	0.3149	0.7455	0.6970	0.6933

Dataset	DE Sen.	DE Spec.	FireFly Sen.	FireFly Spec.	Cuckoo Sen.	Cuckoo Spec.	RBFclassic Sen.	RBFclassic Spec.	ABC Sen.	ABC Spec.	BAT Sen.	BAT Spec.
Blood	0.9908	0.0200	0.9985	0.0250	0.9933	0.0450	0.9795	0.0833	1.0000	0.0050	0.9985	0.0383
Breast	0.7987	0.4877	0.9656	0.9383	0.9707	0.9296	0.9809	0.9259	0.9484	0.9667	0.9777	0.9506
Diabetes	0.2667	0.8205	0.3469	0.8934	0.3052	0.9066	0.5729	0.8795	0.3271	0.9139	0.4260	0.8771
Habitit	0.2300	0.8140	0.4500	0.9209	0.5500	0.9070	0.7000	0.8605	0.3900	0.9395	0.6700	0.8860
Vertebral	0.9267	0.1581	0.8760	0.4516	0.9187	0.3419	0.8933	0.6774	0.8720	0.3742	0.8720	0.5000
Diagnosis I	0.8579	0.5682	0.9789	0.9545	1.0000	0.9636	1.0000	0.7727	0.9526	0.6045	0.9947	0.9636
Diagnosis II	0.8273	0.3947	1.0000	0.9105	1.0000	0.9474	1.0000	1.0000	0.9955	0.8526	1.0000	0.9211
Parkinsons	0.5882	0.4500	0.9706	0.2062	0.9882	0.2562	1.0000	0.2500	0.9569	0.2938	0.9922	0.3812
Liver	0.2160	0.8000	0.4400	0.7265	0.2420	0.8809	0.4600	0.6912	0.1440	0.8750	0.2760	0.8824
Sonar	0.6079	0.4273	0.6000	0.4394	0.3447	0.7697	0.7368	0.8182	0.4395	0.6182	0.4026	0.7485
German	0.4526	0.5429	0.6684	0.3670	0.9522	0.0589	0.8904	0.3839	0.9596	0.0545	0.9689	0.0571
Australian	0.4228	0.5970	0.7554	0.7799	0.7950	0.7940	0.8812	0.8209	0.7604	0.7552	0.8495	0.8224

Table 8.17 The average sensitivity and specificity results of all datasets using different algorithms with (8 Neurons)

Dataset	BBO		GA		PSO		ACO		ES		PBIL	
	Sen.	Spec.	Sen.	Spec.	Sen.	Spec.	Sen.	Spec.	Sen.	Spec.	Sen.	Spec.
Blood	0.9979	0.0483	0.9969	0.0417	0.9995	0.0083	0.9185	0.1133	0.9672	0.0250	0.9851	0.0350
Breast	0.9790	0.9519	0.9720	0.9630	0.9656	0.8654	0.7261	0.3593	0.7191	0.5469	0.9943	0.6802
Diabetes	0.4354	0.8524	0.3063	0.9145	0.3469	0.8651	0.3052	0.7211	0.1927	0.7849	0.1833	0.9398
Habitit	0.6300	0.8744	0.5600	0.9116	0.2100	0.9767	0.4300	0.7116	0.2400	0.7930	0.3300	0.9488
Vertebral	0.8973	0.5097	0.8920	0.4484	0.8773	0.3581	0.7453	0.3226	0.8000	0.2806	0.9000	0.2548
Diagnosis I	1.0000	1.0000	1.0000	1.0000	0.9474	0.6000	0.5000	0.5455	0.5947	0.6091	0.8158	0.6136
Diagnosis II	1.0000	0.9895	1.0000	1.0000	0.9591	0.8000	0.5545	0.4789	0.7773	0.4421	0.8364	0.6526
Parkinsons	0.9922	0.4063	0.9922	0.3438	0.9255	0.2938	0.6549	0.3688	0.5373	0.4188	1.0000	0.0000
Liver	0.3500	0.8824	0.2700	0.9235	0.1880	0.8853	0.3360	0.6559	0.3280	0.6853	0.1580	0.8544
Sonar	0.6816	0.7636	0.1237	0.9879	0.3763	0.6182	0.5947	0.4545	0.3684	0.6545	0.0000	1.0000
German	0.9518	0.2295	0.9662	0.0911	0.5877	0.4902	0.6596	0.3545	0.5592	0.4268	0.9996	0.0009
Australian	0.8634	0.8313	0.8218	0.8470	0.5356	0.6851	0.6307	0.3948	0.5673	0.6090	0.6099	0.7828

Dataset	DE		FireFly		Luckoo		RBFclassic		ABC		BAT	
	Sen.	Spec.	Sen.	Spec.	Sen.	Spec.	Sen.	Spec.	Sen.	Spec.	Sen.	Spec.
Blood	0.9754	0.0400	0.9990	0.0200	0.9985	0.0333	0.9744	0.1000	1.0000	0.0067	0.9974	0.0383
Breast	0.9191	0.7531	0.9732	0.8938	0.9688	0.9556	0.9745	0.9136	0.9675	0.8420	0.9822	0.9593
Diabetes	0.1438	0.9054	0.3104	0.9036	0.3740	0.8813	0.5833	0.8855	0.3542	0.8976	0.4031	0.8825
Habitit	0.2300	0.7791	0.3800	0.9395	0.3900	0.9535	0.7000	0.8372	0.3100	0.9535	0.6100	0.8837
Vertebral	0.8760	0.1710	0.8667	0.4742	0.8827	0.4161	0.8933	0.6774	0.8293	0.4032	0.8813	0.4742
DiagnosisI	0.6053	0.5864	0.9579	0.9364	1.0000	0.9773	1.0000	1.0000	0.8895	0.7136	1.0000	1.0000
DiagnosisII	0.6591	0.6000	0.9955	0.9000	1.0000	0.9263	1.0000	1.0000	0.9318	0.8526	1.0000	0.9842
Parkinsons	0.8725	0.3187	0.9922	0.1688	0.9549	0.3187	0.9412	0.5000	0.9118	0.3563	0.9902	0.2562
Liver	0.2020	0.7941	0.3020	0.7941	0.3420	0.8206	0.4600	0.7059	0.1820	0.8588	0.3800	0.8603
Sonar	0.6132	0.3576	0.5789	0.4424	0.3947	0.7394	0.7632	0.7879	0.4395	0.6697	0.4684	0.7545
German	0.5662	0.4455	0.7925	0.2116	0.8939	0.1321	0.8904	0.3929	0.9513	0.0580	0.9351	0.1027
Australian	0.5356	0.6030	0.6287	0.8239	0.7802	0.7925	0.8713	0.8209	0.6733	0.8060	0.8545	0.8224

Table 8.18 The average sensitivity and specificity results of all datasets using different algorithms with (10 Neurons)

Dataset	BBO Sen.	BBO Spec.	GA Sen.	GA Spec.	PSO Sen.	PSO Spec.	ACO Sen.	ACO Spec.	ES Sen.	ES Spec.	PBIL Sen.	PBIL Spec.
Blood	0.9974	0.0450	0.9995	0.0317	0.9985	0.0150	0.9959	0.0100	0.9513	0.0567	0.9692	0.1083
Breast	0.9809	0.9741	0.9822	0.9358	0.9554	0.8272	0.7076	0.6420	0.8561	0.2333	0.8898	0.7198
Diabetes	0.4625	0.8627	0.3469	0.9042	0.2375	0.8994	0.2104	0.7729	0.1646	0.8723	0.1812	0.9054
Habitit	0.6300	0.8767	0.6300	0.8977	0.3300	0.9395	0.4100	0.5860	0.2100	0.8209	0.1400	0.9767
Vertebral	0.8880	0.5161	0.8933	0.4742	0.8907	0.3355	0.8000	0.1484	0.8347	0.2419	0.8413	0.3581
Diagnosis I	1.0000	0.9591	1.0000	0.9455	0.9000	0.7955	0.5526	0.4909	0.5474	0.5773	0.9789	0.6455
Diagnosis II	1.0000	0.9842	1.0000	1.0000	0.9682	0.7211	0.6727	0.4474	0.6500	0.6632	0.9818	0.6105
Parkinsons	0.9961	0.4437	0.9902	0.3937	0.9098	0.2313	0.6392	0.4000	0.5627	0.4437	1.0000	0.0000
Liver	0.4320	0.8824	0.2420	0.9132	0.1980	0.8515	0.3800	0.6265	0.2720	0.7265	0.2540	0.7750
Sonar	0.6395	0.8515	0.4316	0.8333	0.6526	0.3242	0.6658	0.3333	0.3474	0.6576	0.0000	1.0000
German	0.9268	0.2964	0.9263	0.1768	0.5851	0.4018	0.3789	0.6491	0.6912	0.2973	0.9996	0.0000
Australian	0.8881	0.8231	0.8218	0.8634	0.6980	0.5739	0.6168	0.4627	0.3238	0.7634	0.5257	0.6575

Dataset	DE Sen.	DE Spec.	FireFly Sen.	FireFly Spec.	Luckoo Sen.	Luckoo Spec.	RBFclassic Sen.	RBFclassic Spec.	ABC Sen.	ABC Spec.	BAT Sen.	BAT Spec.
Blood	1.0000	0.0067	0.9944	0.0383	0.9964	0.0283	0.9590	0.3333	0.9964	0.0183	0.9974	0.0550
Breast	0.9497	0.6704	0.9752	0.9481	0.9764	0.9062	0.9554	0.9259	0.9414	0.9111	0.9732	0.9617
Diabetes	0.1281	0.8928	0.3083	0.8892	0.3042	0.8771	0.5417	0.8675	0.3292	0.8801	0.4229	0.8741
Habitit	0.2600	0.9116	0.3300	0.9512	0.4300	0.9163	0.6000	0.8140	0.3000	0.9628	0.6100	0.8767
Vertebral	0.8213	0.2774	0.9013	0.4097	0.8987	0.4129	0.8933	0.6774	0.8853	0.2581	0.8760	0.5419
Diagnosis I	0.6263	0.6364	0.9947	0.9455	1.0000	0.9409	1.0000	0.9545	0.8842	0.8545	1.0000	1.0000
Diagnosis II	0.8000	0.5211	1.0000	0.8947	0.9682	0.9211	1.0000	1.0000	0.9773	0.7579	1.0000	1.0000
Parkinsons	0.5490	0.5375	0.9510	0.3000	0.9941	0.1375	0.9608	0.5000	0.9373	0.2125	0.9843	0.3563
Liver	0.2740	0.7044	0.2800	0.8544	0.2540	0.8515	0.4600	0.7206	0.1900	0.8206	0.3860	0.8412
Sonar	0.5632	0.4273	0.4711	0.5455	0.4737	0.5879	0.7632	0.8485	0.3368	0.6394	0.4184	0.7879
German	0.4899	0.4786	0.8110	0.2098	0.9184	0.0946	0.8684	0.4018	0.9259	0.1036	0.9772	0.0527
Australian	0.5208	0.6664	0.7931	0.7037	0.7267	0.8291	0.8515	0.8358	0.7396	0.7918	0.8525	0.8052

Table 8.19 The Average ranking results obtained by each algorithm in the Friedman test using all datasets

| Algorithm | The number of neurons of hidden layer | | | |
	4	6	8	10
	Ranking			
BBO	2.4167	1.6250	2.4167	2.0000
GA	3.7083	3.6250	3.2500	3.0000
PSO	7.9583	8.3333	7.9167	7.9167
ACO	11.3333	11.0000	11.0833	11.1250
ES	10.8750	11.2500	10.8333	10.8333
PBIL	8.3750	8.1667	8.3758	8.7500
DE	10.2500	10.1667	10.3333	9.9583
FireFly	5.3750	5.7500	6.2083	5.3750
Cuckoo	4.0417	4.7083	4.5833	5.9167
RBFclassic	3.5417	3.3750	3.1250	3.0833
ABC	6.4583	7.0000	6.6667	7.0417
BAT	3.6667	3.0000	3.2083	3.0000

Figure 8.3 shows the complexity of trained RBF network and its corresponding MSE on some of the datasets for each number of neurons in the hidden layer. As shown in all sub-figures for all datasets, BBO has the best results, which has relatively the smallest MSE comparing with all other algorithms except the RBF classic. BBO outperforms all algorithms in term of complexity, which has the smallest complexity values. Moreover, the RBFclassic has the smallest MSE values, but the complexity values are the largest, which outputs complex structure of the RBF network with very low smoothness. The complexity results show the merits of the BBO algorithm in achieving very smooth RBF networks.

Convergence graphs for all algorithms are shown in the Figs. 8.4, 8.5, 8.6 and 8.7 using 4, 6, 8, and 10 neurons, respectively. The convergence curves show the MSE averages of 10 independent runs over 250 iterations. All sub-figures show that BBO is the fastest algorithm in convergence for all datasets. Furthermore, most of other algorithms like DE, ACO, ES, and PBIL have some drawbacks such as trapping at local minima with slow convergence rate. Based on the convergence results, BBO has the superior ability to avoid the local optima.

In summary, the algorithms employed in this work can be classified into three groups: evolutionary, swarm-based, and gradient-based. The results show that evolutionary trainers (including BBO) outperform the other two groups. This is due to the superior local optima avoidance of these algorithms. Evolutionary algorithms mostly have cross-over operators that combine the search agents to create new population(s). Such operators abruptly change the individuals in the population, which results in emphasizing exploration of the search space and local optima avoidance. The gradient-based technique has the least local optima avoidance capability,

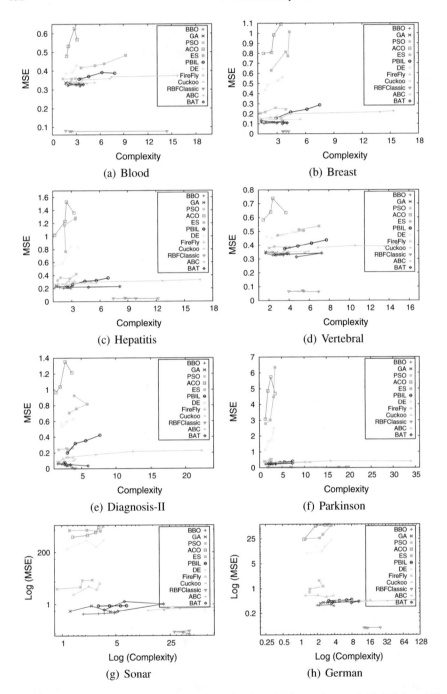

Fig. 8.3 The MSE versus complexity of the classification of different datasets (with 4, 6, 8, and 10 Neurons). **a–i** Results for Blood, Breast, Diabetes, Hepatitis, Vertebral, Diagnosis-I, Diagnosis-II, Parkinson, Liver, Sonar, German, and Australian datasets, respectively

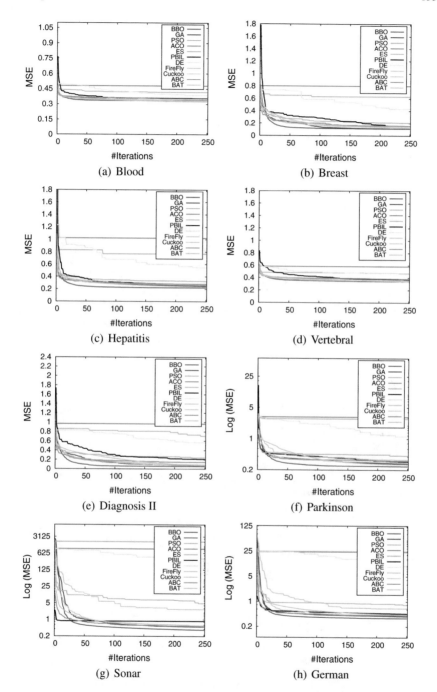

Fig. 8.4 MSE convergence curves of different datasets with (4 Neurons)

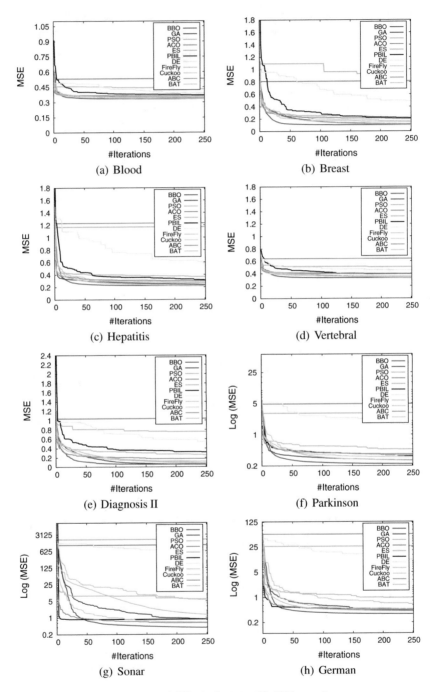

Fig. 8.5 MSE convergence curves of different datasets with (6 Neurons)

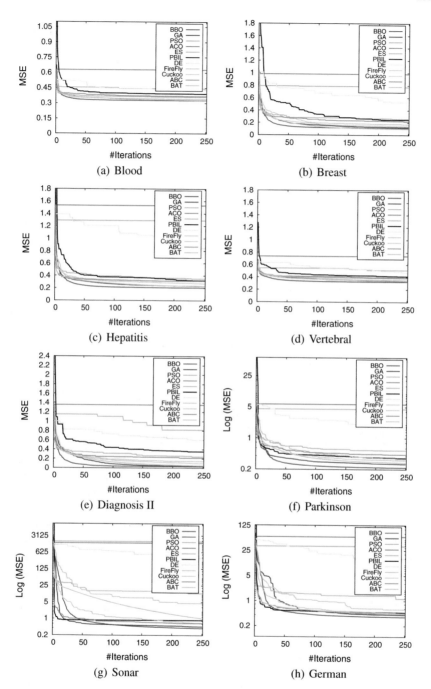

Fig. 8.6 MSE convergence curves of different datasets with (8 Neurons)

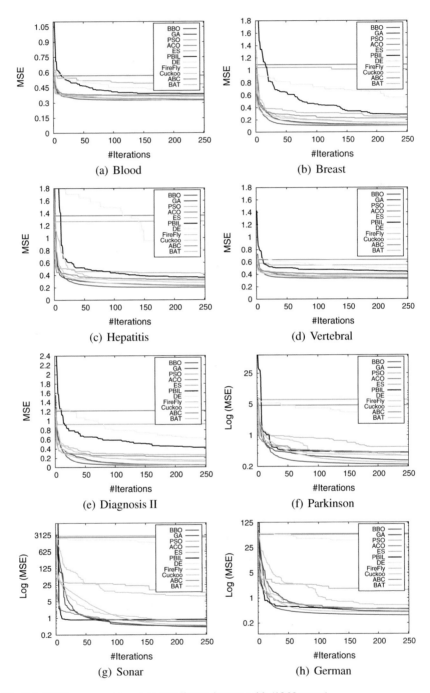

Fig. 8.7 MSE convergence curves of different datasets with (10 Neurons)

which resulted in showing the worse performance on the test cases. The swarm-based algorithms perform better than the gradient-based algorithm because of the higher local optima avoidance. The high local optima avoidance of swarm-based algorithms mostly originate from the population-based nature of these algorithms. However, such algorithms have less intrinsic exploration ability compared to evolutionary algorithms because there are less sudden changes in the search agents.

The results also prove that evolutionary algorithms employed in this work show a better result accuracy and faster convergence rate in average. This shows that high random changes in the search agents of such algorithms does not negatively impact the result accuracy and convergence curve. This originates from the fact that evolutionary algorithms reduce randomness and favor gradual changes with a mechanism called mutation. The mutation operator causes small perturbations and consequently local search around individuals in the population. In other words, the effects of mutation in the overall population is much less than cross-over operators. This operator assists evolutionary algorithms to improve the accuracy of solutions proportional to the number of iterations. Also, the convergence rate is accelerated towards the global optimum by the mutation operator.

Among the swarm-based techniques employed in this work, Cuckoo algorithm outperforms Firefly algorithm, PSO, ABC and ACO. The Cuckoo algorithm has been equipped with a lévy flight which abruptly changes the search agents of this algorithm. Similarly to evolutionary algorithms, this causes extensive exploration of the search space and local optima avoidance significantly. However, other swarm-based algorithm have less operators to promote sudden changes. The ACO algorithm utilizes a pheromone matrix which mostly boosts exploitation and makes this algorithm more suitable for combinatorial problems. The performance of the PSO algorithm also largely depends on the distribution of initial population. This algorithm can easily be trapped in local solution if there is no good distribution in the initial population. The Firefly algorithm also does not random walk or lévy flight, which lead this algorithm to have tendency towards local solutions and less able to avoid them.

In contrary, most of the evolutionary algorithms performed well on the test cases and suppressed swarm-based techniques. Among them, ES and DE showed the poorest performance. In ES, the selection of individual is deterministic which reduces randomness level and exploration of this algorithm. The main operators in this algorithm are mutations which favor exploitation and convergence. These are the main facts that contributed to the failure of this algorithm in solving the datasets. The same statements can be made for DE, but this algorithms has stochastic selection and more crossover operators, which assist it for showing a better performance compared to ES. The performance of the PBIL is better than ES and DE but worse than GA and BBO. This is because PBIL performs crossover on the entire population combined in a vector, which cause better exploration and local optima avoidance compared to ES and DE. However, each individual faces less sudden random changes in comparison with GA and BBO.

BBO outperformed GA because the random changes in the individuals are much higher in this algorithm. The GA algorithm assigns a similar reproduction rate to all the individuals in the population, which causes the same cross-over rate over the course of generations. In contrary, the BBO algorithm assigns each individual a unique emigration and immigration rates. This results in different reproduction rates for each individual and consequently promotion of the exploration and local optima avoidance. Needless to day, this is the main reason of the significant superiority of the BBO-based trainer compared to all trainers employed on all datasets in this work.

The results and discussion of this section show that the BBO algorithm is able to effectively alleviate the drawbacks of the current algorithms when training RBF networks in terms local optima entrapment, result accuracy, and convergence rate.

8.5 Conclusion

This chapter proposed the use of the well-regarded BBO algorithm for training RBF networks to alleviate the drawbacks of conventional and new training algorithms: local optima entrapment, low result accuracy, and slow convergence speed. After proposing the method of training using BBO, it was employed to solve 12 well-known datasets and compared to 11 training methods in the literature including gradient-based, evolutionary, and swarm based algorithms. The algorithms were compared by statistical test on RBF networks with different number of neurons to confidently confirm the performance of the proposed trainer. The results evidently demonstrated that the BBO algorithm is able to outperform the current techniques on the majority of datasets substantially. According to the results, finding, analysis, and discussion of this chapter, the following conclusions can be drawn:

1. BBO shows a fast convergence speed and high result accuracy
2. BBO can avoid local optima in the search space of the training RBF networks problem
3. BBO is able to train RBF networks effectively to classify different datasets with a diverse number of features and training samples
4. BBO is able to train RBF networks with different number of neurons

References

1. Lowe, D., & Broomhead, D. (1988). Multivariable functional interpolation and adaptive networks. *Complex Systems, 2*(3), 321–355.
2. Du, K. L., & Swamy, M. N. (2006). *Neural networks in a softcomputing framework*. Springer Science & Business Media.
3. Hunter, D., Yu, H., Pukish, M. S, I. I. I., Kolbusz, J., & Wilamowski, B. M. (2012). Selection of proper neural network sizes and architecturesa comparative study. *IEEE Transactions on Industrial Informatics, 8*(2), 228–240.
4. Mohaghegi, S., del Valle, Y., Venayagamoorthy, G. K., & Harley, R. G. (2005). A comparison of PSO and backpropagation for training RBF neural networks for identification of a power system with STATCOM. In *Proceedings 2005 IEEE of the Swarm Intelligence Symposium. SIS 2005* (pp. 381–384). IEEE.

5. Leonard, J. A., & Kramer, M. A. (1991). Radial basis function networks for classifying process faults. *IEEE Control Systems, 11*(3), 31–38.
6. Lee, M. J., & Choi, Y. K. (2004). An adaptive neurocontroller using RBFN for robot manipulators. *IEEE Transactions on Industrial Electronics, 51*(3), 711–717.
7. Chng, E. S., Chen, S., & Mulgrew, B. (1996). Gradient radial basis function networks for nonlinear and nonstationary time series prediction. *IEEE transactions on neural networks, 7*(1), 190–194.
8. Castao, A., Fernndez-Navarro, F., Hervs-Martnez, C., Garca, M. M., & Gutirrez, P. A. (2010). Classification by evolutionary generalised radial basis functions. *International Journal of Hybrid Intelligent Systems, 7*(4), 239–248.
9. Schwenker, F., Kestler, H. A., & Palm, G. (2001). Three learning phases for radial-basis-function networks. *Neural Networks, 14*(4–5), 439–458.
10. Wu, Y., Wang, H., Zhang, B., & Du, K. L. (2012). Using radial basis function networks for function approximation and classification. ISRN Applied Mathematics.
11. Mak, M. W., & Cho, K. W. (1998). Genetic evolution of radial basis function centers for pattern classification. In *Proceedings of the IEEE International Joint Conference on Neural Networks, IEEE World Congress on Computational Intelligence* (Vol. 1, pp. 669–673). IEEE.
12. Vakil-Baghmisheh, M. T., & Pave, N. (2004). Training RBF networks with selective backpropagation. *Neurocomputing, 62*, 39–64.
13. Aljarah, I., Faris, H., Mirjalili, S., & Al-Madi, N. (2018). Training radial basis function networks using biogeography-based optimizer. *Neural Computing and Applications, 29*(7), 529–553.
14. Yang, X. S., & Deb, S. (2009). Cuckoo search via Lvy flights. In *World congress on nature & biologically inspired computing*. NaBIC 2009 (pp. 210–214). IEEE.
15. Yang, X. S., & He, X. (2013). Firefly algorithm: Recent advances and applications. *International Journal of Swarm Intelligence, 1*(1), 36–50.
16. Yang, X. S. (2009). Firefly algorithms for multimodal optimization. In *International Symposium on Stochastic Algorithms* (pp. 169–178). Berlin: Springer.
17. Zhang, J., & Sanderson, A. C. (2009). JADE: Adaptive differential evolution with optional external archive. *IEEE Transactions on Evolutionary Computation, 13*(5), 945–958.
18. Mirjalili, S., Mirjalili, S. M., & Lewis, A. (2014). Let a biogeography-based optimizer train your multi-layer perceptron. *Information Sciences, 269*, 188–209.
19. Karaboga, D., & Basturk, B. (2007). A powerful and efficient algorithm for numerical function optimization: Artificial bee colony (ABC) algorithm. *Journal of Global Optimization, 39*(3), 459–471.
20. Yang, X. S. (2010). A new metaheuristic bat-inspired algorithm. In *Nature inspired cooperative strategies for optimization* (NICSO 2010) (pp. 65–74). Berlin: Springer.
21. Simon, D. (2008). Biogeography-based optimization. *IEEE Transactions on Evolutionary Computation, 12*(6), 702–713.
22. Qasem, S. N., Shamsuddin, S. M., & Zain, A. M. (2012). Multi-objective hybrid evolutionary algorithms for radial basis function neural network design. *Knowledge-Based Systems, 27*, 475–497.
23. Asuncion, A., & Newman, D. (2007). UCI machine learning repository.

Chapter 9
Evolutionary Deep Neural Networks

9.1 Introduction

Neural Networks have been widely used to solve a variety of problems in different areas. Such computational techniques mimic the biological interaction between neurons in human brain. The main concepts of NN was first proposed in 1990 [1]. There are different types of NNs: Feed Forward, Radial basis function (RBF), Kohonen self-organizing network [2], and Recurrent neural network [3], The main focus of this paper is on the Feed Forward Neural Networks (FFNs) [4].

In such networks, neurons are arranged into parallel layers and information is exchanged between neuron from one side to another. An FFN with only three layers is called Multi-Layer Perceptron (MLP) [5]. FNNs can be used for function approximation, classification, and data prediction. An FNN with three layers is shown in Fig. 9.1.

The output of the hidden nodes in this network is calculated as follows [4]:

$$s_j = \sum_{i=1}^{n} \left(w_{ij} X_i\right) - \theta_j, \quad j = 1, 2, 3, ..., h \tag{9.1}$$

$$S_j = sigmoid(s_j) = \frac{1}{(1 + e^{-s_j})}, \quad j = 1, 2, 3, ..., h \tag{9.2}$$

where S_j is the output of the j-th hidden node, n is the number of input nodes, θ_j is the bias of jth hidden node, w_{ij} shows the connection weight from ith input node to j-th hidden node, and X_i indicates the ith input.

The output of the network is calculated with the following equations:

$$o_k = \sum_{j=1}^{h} \left(w_{jk} S_j\right) - \theta'_k, \quad k = 1, 2, 3, ..., m \tag{9.3}$$

© Springer International Publishing AG, part of Springer Nature 2019
S. Mirjalili, *Evolutionary Algorithms and Neural Networks*, Studies
in Computational Intelligence 780, https://doi.org/10.1007/978-3-319-93025-1_9

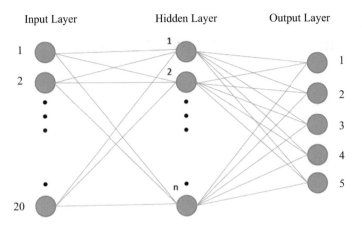

Fig. 9.1 FFN with three layers

$$O_k = sigmoid(o_k) = \frac{1}{(1+e^{-o_k})}, \quad k = 1, 2, 3, ..., m \qquad (9.4)$$

where m is the number of output nodes, S_j is the output of the jth hidden node, θ'_k is the bias of output jth node, and w_{jk} is the connection weight between jth hidden node and k-th output node.

Despite the merits of MLP with only three layers, FNN with more hidden layers have been recently popular as well. They are called Deep Neural Networks (DNN) [6]. The way in which NNs work is the same as that of MLP explained above. However, there are mode hidden layers and the outputs of one hidden layer is calculated based on the previous immediate layer as follows:

$$s_k = \sum_{j=1}^{h} \left(w_{jk} S'_j \right) - \theta'_k, \quad k = 1, 2, 3, ..., m \qquad (9.5)$$

$$S_k = sigmoid(o_k) = \frac{1}{(1+e^{-S_k})}, \quad k = 1, 2, 3, ..., m \qquad (9.6)$$

where S_k is the output of the hidden node, S'_j shows the output of the j-th previous hidden node, and w_{jk} is the connection weight between jth hidden node in the previous later and kth hidden node in the current layer, and m is the number of hidden nodes in the current layer.

In order to classify, predict, or approximate using FNNs, they should be trained with a training algorithm. A training algorithm allows FNNs to learn the dataset and hopefully work with new data. The default training algorithm of FNNs is Back Propagation (BP) [7] algorithm. This algorithm is gradient-based that is able to find the global optimum of the problem of training FNNs. However, it suffers from a major drawback: local optima stagnation [8]. This originates from the gradient-based nature of this algorithm. Since the gradient descent leads a solution towards the steepest slop, this algorithm tends to find local solutions rather than global one [9]. Also, the quality of solution highly depends on the initial solution. This problem deteriorates for DNN due to the large number of parameters to be optimised.

A reliable alternative is to use stochastic optimisation algorithms [10]. Such techniques benefits from high local optima avoidance due to random mechanism [11]. However, they are mostly population-based, so they require more function evaluation and are slow compared to BP and other improved gradient-based algorithms. This area has been very interesting recent. Some of the popular evolutionary algorithm in the field of learning NNs are GA [12], PSO [3, 13], DE [14, 15].

In the following sections, DNNS are trained and applied to a dataset in the field of hand gesture detection.

9.2 Datasets and Evolutionary Deep Neural Networks

9.2.1 Datasets

Unfortunately, there is little standard datasets for hand posture detection as also mentioned in [16]. In order to create the dataset for this work, a software is designed for making different postures with a 3D hand model (see Fig. 9.2). The main model is a 20 Degree of Freedom (DoF) model and has been taken from [17, 18]. It may be seen that there are four sliders for each finger in the software. The user is able to change the orientation of each finger using the four slides. The first two slides control the joint next to the palm in two directions and the rest of sliders are for other joints on the fingers until the top. After getting a desired hand model, the user is able to click on draw model and the software draws the model and give the status of all sliders as a 20-digit number. This software is used to create a dataset with 20 hand poses as can be seen in Fig. 9.3.

The angle of each join for each of the gestures in Fig. 9.3 are presented in Fig. 9.4. This figure also shows that the target vector for each pose. Since there are 20 postures, five bits are allocated to support up to 25 gestures as may be seen in Fig. 9.4.

9.2.2 Evolutionary Neural Networks

In order to design an evolutionary NN, two phases should be done. First, the problem of training NN should be formulated for the evolutionary algorithms. In other words, the problem should be defined in a way to be solved by an optimisation algorithm. Second, a measure should be employed to quantify the performance of NN. This measure will then be used as the objective function to evaluate the search agents of evolutionary algorithms.

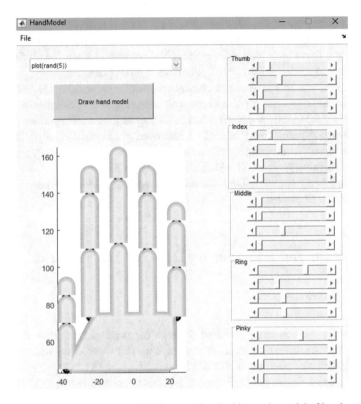

Fig. 9.2 Software developed to create the dataset using the kinematic model of hand

For representing the problem, we have to identify the variables of the problem of training NNs. The main parameters of an NN are connection weights and biases. Therefore, they have to be represented as a vector as follows:

$$\mathbf{x} = \{\overrightarrow{w}, \overrightarrow{b}\} = \{w_1, w_2, ..., b_1, b_2, ...\} \tag{9.7}$$

This vector includes all the parameters to be optimised by a training algorithm. The number of variable in this vector defines the dimension of the search agents (candidate solutions) in evolutionary algorithms.

The next step is to find the objective function. There most widely used measure to quantify the performance of NNs is Mean Square Error (MSE), which is defined as follows:

$$MSE = \sum_{i=1}^{m}(o_i^k - d_i^k) \tag{9.8}$$

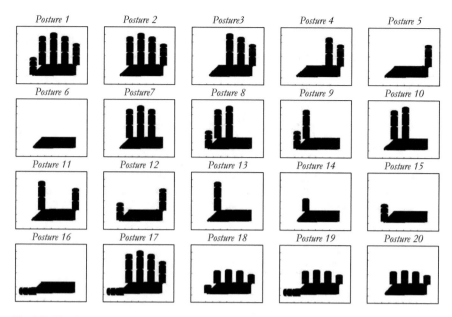

Fig. 9.3 Hand postures used in the software

Posture no	Thumb				Index				Middle				Ring				Pinky				Target				
1	0	0	0	0	0	0	0	0	0	0	0	0	0	0	0	0	0	0	0	0	0	0	0	0	0
2	1	1	0	0	0	0	0	0	0	0	0	0	0	0	0	0	0	0	0	0	0	0	0	0	1
3	0	1	0	0	0	1	0	0	0	0	0	0	0	0	0	0	0	0	0	0	0	0	0	1	0
4	0	1	0	0	0	1	0	0	0	1	0	0	0	0	0	0	0	0	0	0	0	0	0	1	1
5	0	1	0	0	0	1	0	0	0	1	0	0	0	1	0	0	0	0	0	0	0	0	1	0	0
6	0	1	0	0	0	1	0	0	0	1	0	0	0	1	0	0	0	1	0	0	0	0	1	0	1
7	0	1	0	0	0	0	0	0	0	0	0	0	0	0	0	0	0	1	0	0	0	0	1	1	0
8	0	0	0	0	0	0	0	0	0	0	0	0	0	1	0	0	0	1	0	0	0	0	1	1	1
9	0	0	0	0	0	0	0	0	0	1	0	0	0	1	0	0	0	1	0	0	0	1	0	0	0
10	0	1	0	0	0	0	0	0	0	0	0	0	0	1	0	0	0	1	0	0	0	1	0	0	1
11	0	1	0	0	0	0	0	0	0	1	0	0	0	1	0	0	0	0	0	0	0	1	0	1	0
12	0	0	0	0	0	1	0	0	0	1	0	0	0	1	0	0	0	0	0	0	0	1	0	1	1
13	0	1	0	0	0	0	0	0	0	1	0	0	0	1	0	0	0	1	0	0	0	1	1	0	0
14	0	1	0	0	0	0	1	0	0	1	0	0	0	1	0	0	0	1	0	0	0	1	1	0	1
15	0	0	0	0	0	1	0	0	0	1	0	0	0	1	0	0	0	1	0	0	0	1	1	1	0
16	1	0	0	0	0	1	0	0	0	1	0	0	0	1	0	0	0	1	0	0	0	1	1	1	1
17	1	0	0	0	0	0	0	0	0	0	0	0	0	0	0	0	0	0	0	0	1	0	0	0	0
18	0	0	1	0	0	0	1	0	0	0	1	0	0	0	1	0	0	0	1	0	1	0	0	0	1
19	1	0	1	0	0	0	1	0	0	0	1	0	0	0	1	0	0	0	1	0	1	0	0	1	0
20	0	1	0	0	0	0	1	0	0	0	1	0	0	0	1	0	0	0	1	0	1	0	0	1	1

Fig. 9.4 Dataset including 20 samples

where o_i^k is the actual output of the i-th input unit when the k-th training sample appears is training, d_i^k is the desired output of the ith input unit when the kth training sample appears is training, and m is the number of outputs.

The MSE can be calculated for any of the training samples. However, an NN should be adapted to classify all the training samples. It is common in the literature to calculate the MSE for all the training samples and average them. This gives the overall performance of the NN when classifying the training samples. The equation for this purpose is as follows:

$$\overline{MSE} = \sum_{k=1}^{n} \frac{\sum_{i=1}^{m}(o_i^k - d_i^k)}{n} \tag{9.9}$$

where o_i^k is the actual output of the ith input unit when the k-th training sample appears is training, d_i^k is the desired output of the ith input unit when the kth training sample appears is training, m is the number of outputs, and n is the number of training samples.

The problem of training NN can be formulated as follows:

$$Minimise: \ f(\overrightarrow{x}) = \overline{MSE} \tag{9.10}$$

9.3   Results and Discussion

This section covers a series of experiments. First, multiple NNs with different structural parameters are employed to train the datasets. Second, PSO [19] is utilized to train NN for classifying the dataset. Third, different number of hidden nodes, hidden layers, and features are selected to see theirs impacts of the classification. In other words, we investigate the potential of DNNs in recognising hand gestures. PSO [19] is also used to find an optimal set of features.

9.3.1   NN with BP

The first experiment is done with NN and its conventional training algorithm called BP. To see if the NN with the Levenberg-Marquardt (LM) backpropagation algorithm, which is an improved gradient-based BP technique, is effective, the following paragraphs test several NNs with different structures on the dataset. Note that all the NN has 20 inputs and five outputs. However, the number of hidden nodes and hidden layers are different. The training parameters of this algorithm is shown in Table 9.1.

In Table 9.2, an NN with different hidden nodes are trained by the LM algorithm. It may be seen that the best classification rate is 40%. Also, the classification accuracy can be improved up to 15 hidden nodes, but it drops afterwards. This is because a large number of hidden nodes results in over-fitting NN.

Non-linear datasets are better solved by more than one hidden layer in NNs. For this reason, another experiment is conducted with two hidden layers. The results are

Table 9.1 Learning parameters of the LM algorithm

Hidden nodes	Value
Maximum number of epochs to train	1000
Performance goal	0
Maximum validation failures	6
Minimum performance gradient	1.00E−07
Initial mu	0.001
mu decrease factor	0.1
mu increase factor	10
Maximum mu	1.00E+10

Table 9.2 Classification of the dataset using different NNs (20H5) trained by LM

Hidden nodes	Classification accuracy (%)
5	25
10	40
15	30
20	25
30	20

Table 9.3 Classification of the dataset using different NNs (20H1H25) trained by LM

Hidden nodes	Classification accuracy (%)
5 5	40
10 10	55
15 15	50
20 20	30
30 30	20

provided in Table 9.3. Inspecting the results of this table, it is evident the results are consistent with those presented in Table 9.2. The maximum accuracy is again 50% and over-fitting can be seen with 20 hidden nodes, but not in the rest of NNs.

The results of this section show that the dataset created from the kinematic model of hand is very difficult for the conventional training techniques. The next subsection investigates the effectiveness of evolutionary algorithms.

9.3.2 Training FNN Using PSO for the Dataset

In this section, PSO is employed as one of the most well-regarded algorithms in the literature to train FNN for the dataset. A PSO with 60 particles is employed to determine the optimal values for the weights and biases of FNNs over the course of 500 iterations. Since PSO is a stochastic algorithm, we would have to run it 30 times to get reliable results. Two sets of FNN are optimised in this subsection: with one hidden layer and two hidden layers. In both experiments, the number of hidden nodes change. The results are given in Tables 9.4 and 9.5.

Inspecting the results in Table 9.4, it may be seen that the classification accuracy is better than those obtained by LM. This is due to the stochastic nature of PSO and the better local optima avoidance of this algorithm. It is also interesting that over fitting does not occur here. It seems that LM failed to train FNN with a large number of hidden nodes. By contrast, the results of this section show that the PSO is able efficiently find the optimal values or weights and biases of FNN with any number of hidden nodes. The best classification obtained by the PSO-based trainer is 90 for an FNN with 30 hidden nodes.

Table 9.5 shows that despite the large number of connection weights and biases when having two hidden layers, PSO finds trains FNN very well. This results in achieving up to 100% classification rate. It may be seen in Table 9.5 that an FNN with 14 nodes in the first layer and 46 in the second later is the best structure to achieve 100 classification rate. It is worth mentioning here that the training process of a two-layer FNN using PSO takes much longer that FNN with one layer. Therefore, the possibility of reducing the number of hidden nodes and features to achieve less computational time is investigated in the following subsections.

Table 9.4 Classification of the dataset using different NNs (20H5) trained by PSO

Hidden nodes	Classification accuracy	
	Average	Best
5	32	55
10	48	50
15	62.5	65
20	77.67	80
30	82.33	90

Table 9.5 Classification of the dataset using different NNs (20H1H25) trained by PSO

Hidden nodes	Classification accuracy	
	Average	Best
5 5	30	55
10 10	55.5	75
15 15	67	85
20 20	75	90
30 30	85.5	95
14 46	89	100

9.3.3 Number of Hidden Nodes

The preceding section shows that the number of hidden layers and hidden nodes significantly impact the performance of trainers and consequently the classification rate for the datasets. In this experiment, the number of hidden nodes and hidden layers are changed to see how the MSE and classification rates change for the dataset. In order to calculate the average of MSE and classification rate, PSO is run 30 times and collect the results. Also, the best values obtained during the 30 times are provided in this section. The results are illustrated in Fig. 9.5.

It may be seen in this figure that the average MSEs fluctuate, but they tend to decline with increasing the number of iterations. This behaviour is supported by the average classification. Figure 9.5 shows that the average classification rate increases proportional to the number of hidden nodes. The curve peaks at around 40 hidden nodes. These results show that PSO performs stable in training NNs with any number of hidden nodes. The best MSE and classification accuracy are also shown in Fig. 9.5. It is evident in this figure that the curve for best classification rate peaks when the number of hidden nodes is approximately equal to 42. It seems that further increasing the number of hidden nodes results in over-fitting.

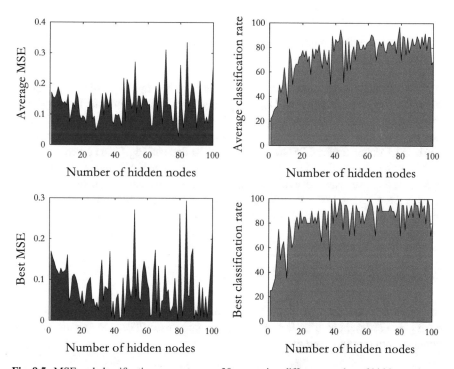

Fig. 9.5 MSE and classification accuracy over 30 runs using different number of hidden nodes.

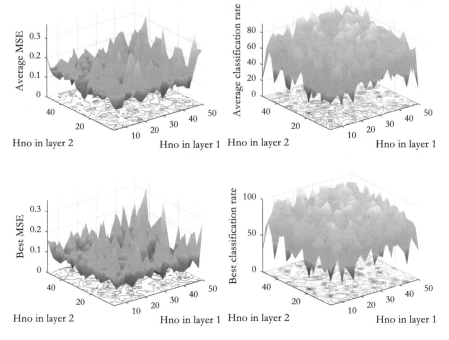

Fig. 9.6 MSE and classification accuracy over 30 runs using different number of hidden nodes.

This subsection also considers the investigating the impacts of different number of hidden nodes in two layers. Due to the high computation cost of training FNN using PSO, we use all the combinations of hidden nodes from 5 to 20 with the step size of 3 (16 * 16 = 256 in total). In order to see the impacts of different hidden nodes on the performance of the PSO-based NN on the dataset, Fig. 9.6 is given.

It may be observed that the results are consistent with those obtained in Fig. 9.5. However, the results are not evidently better, showing that one hidden later is suitable for this dataset. The better classification rate and MSE can be seen, which increase proportional to the number of hidden nodes in both layers.

9.3.4   Changing Nodes in One Layer

In the previous section, we observed that changing the number of nodes in two hidden layers give different results. In order to investigate this further, two experiments are conducted in this subsection:

- Hidden nodes in the first layer change from 1 to 100, and hidden nodes in the second layer equals to 20

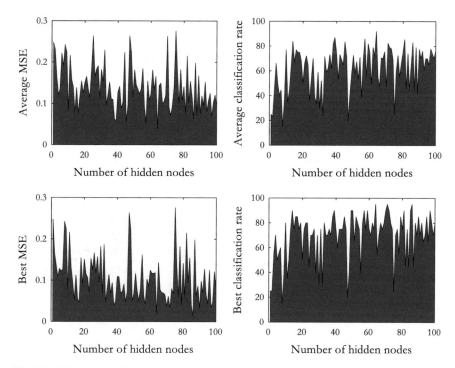

Fig. 9.7 MSE and classification accuracy over 30 runs when the nodes in the first later change but those in the second layer is equal to 20

- Hidden nodes in the first layer equals 20, and hidden nodes in the second layer change from 1 to 100

For each of these combinations, the NN is trained and the results are illustrated in Figs. 9.7 and 9.8.

It may be observed that the results of Fig. 9.8 tend to be better than those in Fig. 9.7. This shows that it seems that seconde hidden later plays a more important role in the final classification rate. It is worth noting that 100% classification rate is achieved when the structure of NN is 20 * 65 and 20 * 82.

9.3.5 Changing the Number of Layers (Deep Learning)

Another experiment conducted in this chapter is to find out the impact of the number of hidden layers on the performance of learning algorithm and NNs for the datasets that we have created. Due to the expensive computational cost of DNNs, 20 or 50

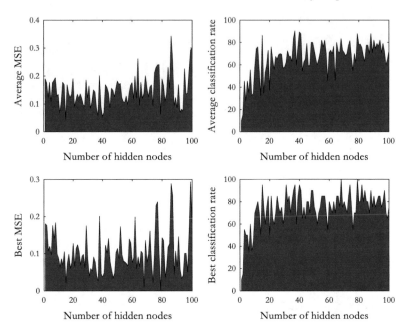

Fig. 9.8 MSE and classification accuracy over 30 runs when the nodes in the second layer change but those in the first layer is equal to 20

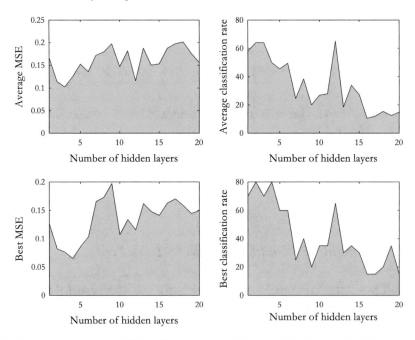

Fig. 9.9 Average and best MSE/classification rate in different DNN with only 20 hidden nodes in each layer

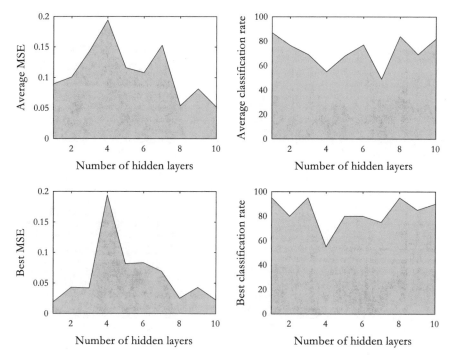

Fig. 9.10 Average and best MSE/classification rate in different DNN with only 50 hidden nodes in each layer

nodes are considered in two independent experiments, and the number of hidden layers is only changed from 1 to 20 or 1 to 10. The results are illustrated in Figs. 9.9 and 9.10.

Surprisingly, Figs. 9.9 and 9.10 show that DNN with 20 and 50 nodes show different behaviours. It seems that none of the DNNs provide good results when there are 20 nodes. In fact, the performance significantly degrades when there are only 20 nodes in all layers. However, the performance is almost stable when we have 50 hidden nodes in each layer. This shows that increasing the number of hidden layers does not necessarily increase the classification rate. Two to four hidden layers suffice to get a reasonable classification rate.

9.3.6   Feature Selection

In the last subsection, a method is used to select an optimal set of features on the dataset to see if the computational cost of the trainer can be reduced. There are two methods for selecting features: wrapper versus filters [20]. In the former method, a set of best features is chosen considering an objective function. In the latter approach,

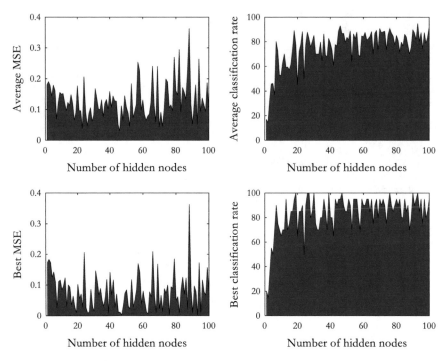

Fig. 9.11 MSE and classification accuracy over 30 runs using different number of hidden nodes and only 11 features selected by PSO.

however, the best features are selected only based on the dataset itself. A BPSO-based wrapper in this subsection is employed to find the optimal set of features. The BPSO is then applied to an FNN with one hidden layer and 40 nodes to find the optimal set of features. This algorithm finds 11 features out of 20. The 11 obtained features are then to train different NNs and find the average/best MSE and classification rates. The results are shown in Fig. 9.11.

With the comparison of Figs. 9.5 and 9.11, it may be seen that the MSE and classification rates are much better with the features selected by BPSO. It is interesting that 100% classification rate occurs around 43 hidden nodes when using all the features. However, 100% classification rate is achieved by a FNN with only 20 hidden nodes when using 11 features optimised. This evidently shows that feature selection using BPSO was effective and results in less computational cost when training FNN since we need fewer connection weights and biases to be optimised.

9.4 Conclusion

This work comprehensively investigated the application of NNs and EAs in the field of hand posture/gesture detection. Due to the little number of datasets when using a kinematic model of hand, we developed a software to create the dataset. After making the dataset, different NNs trained with gradient-based and evolutionary trainers were employed to classify the dataset. Several experiments were done on the NNs. First, an improved BP was compared to PSO and it was observed that PSO is able to outperform the gradient-based trainers. The superiority was significant when there was a large number of hidden node, which is due to the local optima stagnation of BP algorithms. Secondly, the number of hidden nodes in different layers were changed to see their impacts on the performance of the trainer. It was observed that the large number of hidden nodes results in overfitting. The chapter also reached 100% classification accuracy with changing the hidden nodes. Thirdly, since it was difficult to find out which hidden layer is more important for the dataset, another experiment was conducted by changing the hidden nodes in only one of the hidden layers. It was observed that the second layer is more important that the first one.

Fourthly, the impact of different hidden layers on the performance of trainers when training DNNs was investigated. It was observed that varying the number of hidden layers does not significantly increase the performance and there should be a large number of hidden nodes (50) to get reasonable results. Finally, BPSO was employed to find the optimal set of features. BPSO managed to find 11 features, which resulted in decreasing the complexity of the dataset and 100% classification was achieved with approximately half the number of hidden nodes that needed when using all features.

References

1. McCulloch, W. S., & Pitts, W. (1990). A logical calculus of the ideas immanent in nervous activity. *Bulletin of Mathematical Biology, 52*(1), 99–115.
2. Kiang, M. Y. (2001). Extending the Kohonen self-organizing map networks for clustering analysis. *Computational Statistics & Data Analysis, 38*, 161–180.
3. Gudise, V. G., & Venayagamoorthy, G. K. (2003) Comparison of particle swarm optimization and backpropagation as training algorithms for neural networks. In *Proceedings of the 2003 IEEE swarm intelligence symposium*. SIS'03 (pp. 110–117).
4. Bebis, G., & Georgiopoulos, M. (1994). Feed-forward neural networks. *IEEE Potentials, 13,* 27–31.
5. Lang, B. (2005). Monotonic multi-layer perceptron networks as universal approximators. In *International conference on artificial neural networks* (pp. 31–37).
6. Ciregan, D., Meier, U., & Schmidhuber, J. (2012). Multi-column deep neural networks for image classification. In *2012 IEEE conference on computer vision and pattern recognition (CVPR)* (pp. 3642–3649).
7. Li, J., Cheng, J.-H., Shi, J.-Y., & Huang, F. (2012). Brief introduction of back propagation (BP) neural network algorithm and its improvement. In *Advances in computer science and information engineering* (pp. 553–558). Springer.

8. Lee, Y., Oh, S.-H., & Kim, M. W. (1993). An analysis of premature saturation in back propa-
    gation learning. *Neural Networks, 6,* 719–728.
9. Ibnkahla, M. (2003). Nonlinear system identification using neural networks trained with natural
    gradient descent. *EURASIP Journal on Advances in Signal Processing, 2003,* 1–9.
10. Siddique, N., & Adeli, H. (2013). Evolutionary neural networks. In Computational intelligence:
    synergies of fuzzy logic, neural networks and evolutionary computing (pp. 307–355).
11. BoussaD, I., Lepagnot, J., & Siarry, P. (2013). A survey on optimization metaheuristics. *Infor-
    mation Sciences, 237,* 82–117.
12. Schaffer, J. D., Whitley, D., Eshelman, L. J. (1992) Combinations of genetic algorithms and
    neural networks: A survey of the state of the art. In *International workshop on combinations
    of genetic algorithms and neural networks.* COGANN-92 (pp. 1–37).
13. Khan, K., & Sahai, A. (2012). A comparison of BA, GA, PSO, BP and LM for training feed
    forward neural networks in e-learning context. *International Journal of Intelligent Systems and
    Applications, 4,* 23.
14. Ilonen, J., Kamarainen, J.-K., & Lampinen, J. (2003). Differential evolution training algorithm
    for feed-forward neural networks. *Neural Processing Letters, 17,* 93–105.
15. Magoulas, G. D., Plagianakos, V. P., & Vrahatis, M. N. (2004). Neural network-based colono-
    scopic diagnosis using on-line learning and differential evolution. *Applied Soft Computing, 4,*
    369–379.
16. Oberweger, M., Riegler, G., Wohlhart, P., & Lepetit, V. (2016). Efficiently creating 3D training
    data for fine hand pose estimation. arXiv preprint arXiv:1605.03389.
17. Malvezzi, M., Gioioso, G., Salvietti, G., Prattichizzo, D., & Bicchi, A. (2013). Syngrasp: A
    matlab toolbox for grasp analysis of human and robotic hands. In *IEEE international conference
    on robotics and automation (ICRA)* (pp. 1088–1093).
18. Malvezzi, M., Gioioso, G., Salvietti, G., & Prattichizzo, D. (2015). Syngrasp: A matlab toolbox
    for underactuated and compliant hands. *IEEE Robotics & Automation Magazine, 22,* 52–68.
19. Kennedy, J. (2011). Particle swarm optimization. In *Encyclopedia of machine learning* (pp.
    760–766). Springer.
20. Das, S. (2001). Filters, wrappers and a boosting-based hybrid for feature selection. In *ICML*
    (pp. 74–81).

Printed in the United States
By Bookmasters